THE MIDDLE EAST AND NORTH AFRICA

The global political map is undergoing a process of rapid change as former states disintegrate and new states emerge. Territorial change in the form of conflict over land and maritime boundaries is inevitable but the negotiation and management of these changes threaten world peace.

The Middle East and North Africa brings together some of today's most influential analysts on a region which from colonial times to the present has seen great territorial change.

World Boundaries is a unique series embracing the theory and practice of boundary delimitation and management, boundary disputes and conflict resolution, and territorial change in the new world order. Each of the five volumes – *The Middle East and North Africa, Eurasia, The Americas, Maritime Boundaries* and *Global Boundaries* – is clearly illustrated with maps and diagrams and contains regional case-studies to support thematic chapters. This series will lead to a better understanding of the means available for the patient negotiation and peaceful management of international boundaries.

Clive H. Schofield is Executive and Research Officer at the International Boundaries Research Unit at the University of Durham, and *Richard N. Schofield* is Deputy Director of the Geopolitics Research Centre, School of Oriental and African Affairs, University of London.

WORLD BOUNDARIES SERIES
Edited by Gerald H. Blake
*Director of the International Boundaries Research Unit
at the University of Durham*

The titles in the series are:

GLOBAL BOUNDARIES
Edited by Clive H. Schofield

THE MIDDLE EAST and NORTH AFRICA
Edited by Clive H. Schofield and Richard N. Schofield

EURASIA
Edited by Carl Grundy-Warr

THE AMERICAS
Edited by Pascal Girot

MARITIME BOUNDARIES
Edited by Gerald H. Blake

THE MIDDLE EAST AND NORTH AFRICA

World Boundaries Volume 2

Edited by Clive H. Schofield and Richard N. Schofield

London and New York

First published 1994
by Routledge
2 Park Square, Milton Park, Abingdon, Oxon, OX14 4RN

Simultaneously published in the USA and Canada
by Routledge
270 Madison Ave, New York NY 10016

Reprinted 1997

Transferred to Digital Printing 2007

© 1994 International Boundaries Research Unit

Typeset in 10pt September by Solidus (Bristol) Limited

British Library Cataloguing in Publication Data
A catalogue record for this book is available from the British Library

Library of Congress Cataloging in Publication Data
Has been applied for

ISBN 0–415–08839–9
5-vol. set: ISBN 0–415–08840–2

Publisher's Note
The publisher has gone to great lengths to ensure the quality
of this reprint but points out that some imperfections in
the original may be apparent

CONTENTS

List of plates	vii
List of figures	viii
List of tables	x
List of contributors	xi
Series preface	xiii
Preface	xv
Acknowledgements	xx

1 TERRITORY, STATE AND NATION IN THE MIDDLE EAST AND NORTH AFRICA
George Joffé — 1

2 TRANSBOUNDARY INTERACTION AND POLITICAL CONFLICT IN THE CENTRAL MIDDLE EAST
The case of Syria
Alasdair D. Drysdale — 21

3 DEMARCATION LINES IN CONTEMPORARY BEIRUT
Michael F. Davie — 35

4 THE PALESTINIAN STATE
Options and possibilities
Ghada Karmi — 59

5 THE FUNCTIONAL PRESENCE OF AN 'ERASED' BOUNDARY
The re-emergence of the Green Line
David Newman — 71

CONTENTS

6 RIVER AND LAKE BOUNDARIES IN ISRAEL
   Gideon Biger   99

7 GAZA VIABILITY
   The need for enlargement of its land base
   Saul B. Cohen   110

8 THE EVOLUTION OF THE
   TRANSJORDAN–IRAQ BOUNDARY, 1915–40
   Vartan Amadouny   128

9 LEGAL ASPECTS OF THE IRAQI
   SOVEREIGNTY AND BOUNDARY DISPUTES
   WITH KUWAIT
   Maurice H. Mendelson   142

10 THE HISTORICAL PROBLEM OF IRAQI
   ACCESS TO THE PERSIAN GULF
   The interrelationships of territorial disputes with Iran and
   Kuwait, 1938–90
   Richard N. Schofield   158

11 THE UNITED ARAB EMIRATES AND OMAN
   FRONTIERS
   Julian Walker   173

12 CAPTAIN KELLY AND THE
   SUDAN–UGANDA BOUNDARY
   COMMISSION OF 1913
   Gerald H. Blake   184

   Index   197

PLATES

12.1 Captain Harry Kelly, RE, killed in action in France, 24 October
1914   185
12.2 Members of the Boundary Commission with District
Commissioners at Nimule, January 1913; Kelly is seated second
from left, Tufnell on the right   188
12.3 Kelly described these men as his 'intelligence staff', including
Adam with the perambulator wheel   189

FIGURES

2.1 Lebanese arrivals in Syria   23
2.2 Syrian departures to Lebanon   24
2.3 Jordanian arrivals in Syria   27
2.4 Syrian–Jordanian trade   29
2.5 Iraqi arrivals in Syria   30
2.6 Iranian arrivals in Syria   32
3.1 Early nineteenth-century Beirut with location of the
Greek Orthodox quarter   36
3.2 Distribution of religious communities in pre-1975 Beirut   44
3.3 Spatial distribution of communities in Moussaytbeh
(West Beirut)   46
3.4 Spatial distribution of communities: the situation
in 1990   49
3.5 Inter-territorial relations before 1975 (A), and during
1975–90 (B)   54
5.1 Formal and functional characteristics of boundaries:
the feedback effect   72
5.2 The 'Green Line' as a boundary of 'price discontinuity'   79
5.3 Municipal affiliation of Jewish settlements either side of
the 'Green Line'   81
5.4 New settlement planning along the 'Green Line'   83
5.5 The 'Green Line' as a barrier: travelling alternatives to
Jerusalem   86
5.6 Road sign at Shoqet Junction showing optional routes to
Jerusalem   88
5.7 The 'Green Line' as depicted in cartoons   89
5.8 Possible territorial exchanges: re-demarcation of the
'Green Line'   93

5.9 Jewish settlements as wedges in the way of territorial
    exchange   95
6.1 The boundary between Huleh (Hula) and the Sea of Galilee   100
6.2 Armistice line: Israel–Syria   103
6.3 Armistice line: Israel–Jordan   105
7.1 A Gaza mini-state   116
8.1 The Iraq–Jordan boundary   129
8.2 The Jordan–Syria boundary   131
8.3 The War Office map of Asia, 1918   137
9.1 Red and Green lines introduced to define Kuwaiti territory
    after the Anglo-Ottoman Convention, 29 July 1913   145
9.2 Disagreement between the Foreign Office and the
    Government of India during 1941 and 1942 on where
    the land boundary should terminate on the Khor Zubair   154
10.1 Iraq's limited access to the sea   159
12.1 The political context of the Sudan–Uganda Boundary
     Commission in 1913   187
12.2 The Sudan–Uganda boundary today   194

TABLES

4.1 Palestinian populations in the Middle East, 1989   60
5.1 Phases of 'Green Line' boundary   75
7.1 Projected land requirements for Gaza state   115

NOTES ON CONTRIBUTORS

Vartan Amadouny, Department of Geography, University of Southampton, UK

Gideon Biger, Department of Geography, Tel Aviv University, Tel Aviv, Israel

Gerald H. Blake, Director, International Boundaries Research Unit, University of Durham, UK

Saul B. Cohen, Department of Geology and Geography, Hunter College, City University, New York, USA

Michael F. Davie, Université François Rabelais, Tours, France

Alasdair D. Drysdale, Department of Geography, University of New Hampshire, Durham, USA

George Joffé, Geopolitics Research Centre, School of Oriental and African Studies, University of London, UK

Ghada Karmi, School of Oriental and African Studies, University of London, UK

Maurice H. Mendelson, Professor of International Law, University of London, UK

David Newman, Department of Geography, Ben Gurion University, Beer Sheva, Israel

Richard N. Schofield, Geopolitics Research Centre, School of Oriental and African Studies, University of London, UK

Julian Walker, Foreign and Commonwealth Office, UK

SERIES PREFACE

The International Boundaries Research Unit (IBRU) was founded at the University of Durham in January 1989, initially funded by the generosity of Archive Research Ltd of Farnham Common. The aims of the unit are the collection, analysis and documentation of information on international land and maritime boundaries to enhance the means available for the peaceful resolution of conflict and international transboundary cooperation. IBRU is currently creating a database on international boundaries with a major grant from the Leverhulme Trust. The unit publishes a quarterly *Boundary and Security Bulletin* and a series of *Boundary and Territory Briefings*.

IBRU's first international conference was held in Durham on September 14–17, 1989 under the title of 'International Boundaries and Boundary Conflict Resolution'. The 1989 Conference proceedings were published by IBRU in 1990 edited by C.E.R. Grundy-Warr: *International Boundaries and Boundary Conflict Resolution*. The theme chosen for our second conference was 'International Boundaries: Fresh Perspectives'. The aim was to gather together international boundary specialists from a variety of disciplines and backgrounds to examine the rapidly changing political map of the world, new technical and methodological approaches to boundary delimitations, and fresh legal perspectives. Over 130 people attended the conference from 30 states. The papers presented comprise four of the five volumes in this series (Volumes 1–3 and Volume 5). Volume 4 largely comprises proceedings of the Second International Conference on Boundaries in Iberocamerica held at San Jose, Costa Rica, November 14–17, 1990. These papers, many of which have been translated from Spanish, seemed to complement the IBRU conference papers so well that it was decided to ask Dr Pascal Girot, who is coordinator of a major project on border regions in Central America based at CSUCA (The Confederation of Central

American Universities), to edit them for inclusion in the series. Volume 4 is thus symbolic of the practical cooperation which IBRU is developing with a number of institutions overseas whose objectives are much the same as IBRU's. The titles in the *World Boundaries* series are:

Volume 1 *Global Boundaries*
Volume 2 *The Middle East and North Africa*
Volume 3 *Eurasia*
Volume 4 *The Americas*
Volume 5 *Maritime Boundaries*

The papers presented at the 1991 IBRU conference in Durham were not specifically commissioned with a five-volume series in mind. The papers have been arranged in this way for the convenience of those who are most concerned with specific regions or themes in international boundary studies. Nevertheless, the editors wish to stress the importance of seeing the collection of papers as a whole. Together they demonstrate the ongoing importance of research into international boundaries on land and sea, how they are delimited, how they can be made to function peacefully, and perhaps above all how they change through time. If there is a single message from this impressive collection of papers it is perhaps that boundary and territorial changes are to be expected, and that there are many ways of managing these changes without resort to violence. Gatherings of specialists such as those at Durham in July 1991 and at San Jose in November 1990 can contribute much to our understanding of the means available for the peaceful management of international boundaries. We commend these volumes as being worthy of serious attention, not just from the growing international community of border scholars but from decision-makers who have the power to choose between patient negotiation and conflict over questions of territorial delimitation.

Gerald H. Blake
Director of IBRU
Durham, January 1993

PREFACE

Geographers, lawyers and practitioners all contributed what at first sight seemed an eclectic array of papers dealing with Middle Eastern questions at IBRU's second conference – 'International Boundaries: Fresh Perspectives', during July 1991. Yet this was an exciting time for territorial issues in the region. The Iraqi invasion of Kuwait just less than a year earlier had raised awkward questions about the nature of the territorial state in the region. The United Nations-sponsored settlement of the subsequent Gulf War in the spring of 1991 saw the Secretary-General of that organisation make the decision to intervene directly in the historically explosive Iraq–Kuwait border question by taking responsibility for the demarcation and final settlement of this troublesome territorial limit. There also seemed a better opportunity than ever for the West to address seriously the Palestinian problem. Partially by design, partially by accident, half of the twelve papers presented here deal, on the one hand, with geographical/historical and legal aspects of Iraq–Kuwait territorial and sovereignty disputes and, on the other hand, with territorial dimensions of the Palestinian question.

Many of the contributions were historical in nature, dealing with aspects relating to the creation of territorial limits in the region. At least three studies testify to the vagaries of the boundary delimitation process during the colonial period, which certainly, in the case of Iraq–Kuwait, has not helped the United Nations in its present quest to finalise this territorial limit. Others utilise little-known source material. One contribution gives the perspective of a boundary-maker. Another looks at the history of demarcation lines within a strife-torn urban system. The interrelationship of the border disputes of one state with a narrow land corridor to the sea with its two neighbours in a highly volatile geostrategic region is also reviewed. The last of this historical group reviews the changing functional importance of a territorial limit over time.

Two geographical studies break this historical mould. One traces the effect of political conflict upon transboundary interaction. Another imaginatively proposes a geographical solution to one of the region's most entrenched territorial and political problems.

George Joffé provides a most useful overview of the relationship between territory, state and nation in Chapter 1 that serves as the perfect backcloth to many of the subsequent contributions dealing with historical aspects of boundary delimitation. The author's tried and tested theme of the conflict between European and indigenous modes of social and spatial organisation is lucidly articulated.

Chapter 2 is contributed by Alasdair Drysdale, a political geographer and a respected Syrian specialist. He combines these strengths to trace how political conflict has affected the status of Syria's boundaries, namely its permeability/impermeability to flows of trade and people. For Syria juridical disputes over boundary delimitation have not been a feature of inter-state relations but functional disputes most certainly have been. The propensity of Syria and its neighbours to close boundaries in times of political dispute has seriously fractured those patterns of cross-boundary interaction that have had the chance to develop. For the sake of regional development and cooperation, it would clearly be useful if political dispute could find more constructive expression in the future than the temporary closure of international boundaries.

The concept and reality of spatial division within Beirut, concentrating on the last two decades, is the subject of Michael Davie's fascinating chapter. He does not restrict himself to the city's major boundary between the Muslim west and the Christian east (the first notorious Green Line to be reviewed in this volume), but instead follows the history and evolution of a multiplicity of visible and invisible *de facto* lines which have based themselves on various discontinuities, be they ethnic, linguistic, religious or social. The interrelationships between the small territories bounded by these limits amidst the breakdown of the traditional urban system in Beirut, are perceptively identified.

In Chapter 4 Ghada Karmi assumes, like many other informed observers (Palestinian, Israeli and others), that the best hope for a lasting peace in the Arab–Israeli conflict lies in the adoption of the two-state option. After clearly enumerating the principles which Palestinians demand to see respected en route to statehood, Karmi then spells out the issues which need to be broached in the institution of the two-state system. The chapter ends with commentaries on differing plans for the two-state options put forward by Israelis and Palestinians.

David Newman presents an immaculately documented and illustrated functional analysis of Israel's Green Line in Chapter 5. The way in which this territorial limit is perceived by inhabitants on either side, within Israel or the West Bank, is radically different, while both the functional and physical importance of the divide has changed over time. If the 'Peace Process' ultimately results in the two-state option the Green Line will become, for the first time in its existence, a recognised international boundary. If the Palestinians arrive at something less than full statehood, the territorial limit will still represent a major discontinuity. Newman argues persuasively that whatever the formal status of the Green Line in the future, discontinuities must not be reinforced by the sealing of boundaries – the functional characteristics of the limit must tend towards contact, at least in the economic sphere.

Chapter 6 has been contributed by Gideon Biger, a historical geographer with an intense interest in the way Britain shaped the territorial limits of Palestine early in this century. In this piece Biger points to the lack of care Britain took when dealing with water bodies in initially delimiting the eastern boundaries of Palestine. This will hold serious implications, as Biger suggests, when and if Israel and its eastern Arab neighbours come to a final political settlement and elaborate further territorial definition.

It is a great privilege for IBRU that Chapter 7 was contributed by one of political geography's great thinkers. Thirty years on from his landmark *Geography and Politics in a World Divided* (1963), Saul Cohen has produced another highly inventive and imaginative geographical solution to a persistently hopeless political and territorial problem. By advocating an expansion of Gaza into the Egyptian Sinai, Cohen reckons the strip might attain 'viable mini state' status. Certainly the present future of Gaza within a non-contiguous Palestinian state or entity seems fraught with potential difficulties. Geographers can always come up with solutions such as the creation of land corridors, or, in the right circumstances, exchanges or movements of populations. Ultimately it is left to the politicians to act upon them, however.

Vartan Amadouny gives a well-researched account of the way in which the British government stumbled towards a definition of the Transjordan–Iraq boundary in Chapter 8. Contradictory political commitments and definitions of boundary points coupled with inaccurate cartography make it apparent that British policy towards this territorial limit at least, was far from concerted. It is a story which is not restricted to just one boundary in the region, as we shall discover.

Maurice Mendelson covers an immense amount of ground in a short space of time in Chapter 9 when reviewing Iraq's claims to the entirety of Kuwait and also to northern Kuwait and its islands. In what is a commendably concise account for a lawyer, Mendelson has little difficulty invalidating the Iraqi historical claims to sovereignty and quickly identifies (remember that this was written in July 1991) the problems with the vague Iraq–Kuwait land boundary delimitation which have subsequently troubled the UN team charged with responsibility for its demarcation.

In Chapter 10 Richard Schofield highlights the consistently fraught geopolitical problems of Iraqi access to the Gulf by exploring the inter-relationship of the Baghdad government's territorial disputes with Iran over the Shatt al-Arab and Kuwait. He concludes that it is vital for the future security of the northern Gulf that in the medium to long term Iraq no longer perceives itself as 'squeezed' out of the Gulf. Just how its persistent negative consciousness should be redressed is the problem that remains to be answered.

It is always refreshing to hear the view of a practitioner on boundary questions rather than those of an academic who takes a retrospective interest in a particular case-study, perhaps as the result of visiting the area in question or coming across apparently significant materials in government archives. This is particularly so when the practitioner in question actually bore a considerable responsibility for the boundary in question. Julian Walker has never been very far away from any major territorial settlement or dispute in the Persian Gulf region for the last four decades or so. Recently he has played an important advisory role in the UN-sponsored settlement of the Iraq–Kuwait border question. He is perhaps best known, however, for shaping the patchwork quilt of non-contiguous and enclaved sheikhdoms which today comprise the United Arab Emirates. In Chapter 11 Walker reflects upon his contribution towards solving territorial disputes between the component sheikhdoms of the former Trucial Coast and also upon the punctuated border he helped bring into being to the south with the Sultanate of Muscat and Oman. The cluttered contemporary appearance of these territorial limits arises from Britain's only effort in the peninsula to settle borders primarily according to tribal factors. Though untidy, Walker regards them to have served the purpose for which they were drawn.

The last chapter in the volume is contributed by IBRU Director Gerald Blake and makes interesting use of archival sources existing within Durham itself (the Sudan Archive) to tell the story of the initial drawing of the Sudan's southern boundaries on the eve of the Great

War. Central to the whole business was a certain Captain Kelly, an unsung colonial boundary maker, certainly not as well known as those celebrated 'champions' of the natural boundary, Lord Curzon and Colonel Holdich of the government of India. Perhaps with the forthcoming release of his edited collection of Captain Kelly's diaries relating to Sudanese boundary matters, Gerald will be able to change all of this.

ACKNOWLEDGEMENTS

Much of the initial work on these proceedings was undertaken by IBRU's executive officer Carl Grundy-Warr before his appointment to the National University of Singapore early in 1992. It has taken a team of editors to complete the task he began so well. Elizabeth Pearson and Margaret Bell assisted in the preparation of the manuscripts for several of these volumes, and we acknowledge their considerable contribution. John Dewdney came to our rescue when difficult editorial work had to be done. In addition many people assisted with the organisation of the 1991 conference, especially Carl Grundy-Warr, Greg Englefield, Clive Schofield, Ewan Anderson, William Hildesley, Michael Ridge, Chng Kin Noi and Yongqiang Zong. Their hard work is gratefully acknowledged. We are most grateful to Tristan Palmer and his colleagues at Routledge for their patience and assistance in publishing these proceedings, and to Arthur Corner and his colleagues in the Cartography Unit, Department of Geography, University of Durham for redrawing most of the maps.

Gerald H. Blake
Director, IBRU

1

TERRITORY, STATE AND NATION IN THE MIDDLE EAST AND NORTH AFRICA

George Joffé

INTRODUCTION

The legal attributes of the modern state have been well defined over the past two centuries in customary international law and by enactments of international bodies such as the United Nations. These definitions, however, stem from legal principles and political concepts which evolved in Europe (Akehurst 1987: 20–3) rather than from the experience of state systems elsewhere in the world.

THE DEFINITION OF THE SOVEREIGN STATE

One of the most important attributes of the state under international law has been sovereignty – the absolute right of the ruling entity of the state to exercise power within it. Since the Peace of Westphalia, states have been considered sovereign entities precisely in this sense (Thomas 1985: 3). Crucial to such views, furthermore, has been the principle that sovereignty is linked to territory, since 'Territory is a tangible attribute of statehood and within that particular geographical area which it occupies, a state enjoys and exercises sovereignty' (Wallace 1986: 81).

A state has, therefore, the right of exclusive sovereign jurisdiction over the territory it controls, for the most commonly accepted definition of state sovereignty is territorial sovereignty (Wallace 1986: 102). In the classic definition of that exclusive jurisdiction, it has, 'the right to exercise therein (in its territory), to the exclusion of any other state, the function of a state' (*Island of Palmas* case (1928), UNRIAA II 829, 838). It follows from this statement that the precise determination of the territorial extent of a state is crucially important in determining the extent of its jurisdiction – although other factors may affect this, both in terms of limitations and in terms of applicability (Wallace 1986: 101–6).

Equally significant is the associated legal concept of the state itself. In international law, this also has a specific definition. States are entities with populations living on territories effectively controlled by governments which are also capable of conducting international relations with other states (Akehurst 1987: 53). Interestingly enough, this definition does not require that the precise borders of the states in question are definitively established or are undisputed. It does, however, imply that there is, in principle, a definable territorial extent to the state which would, thereby, acquire precise territorial limits. The definition also implies that a state does not necessarily cease to exist if it is occupied by an act of war, provided other states dispute that act of conquest (Akehurst 1987: 53) and, thereby, implicitly continue the practice of conducting relations with it. Indeed, the recent conflict over Iraq's occupation and annexation of Kuwait was implicitly based on just such a principle of international law, even though it may have been explicitly justified by recourse to United Nations Security Council resolutions or chapters of the United Nations Charter.

THE STATE IN ISLAM

Although these definitions might seem self-evident within the context of the modern international order, this has not necessarily been the case in the Middle East or North Africa where alternative concepts of the state have traditionally been recognised. In fact, there is a small corpus of Islamic constitutional law that has defined an alternative model for political entities within the Islamic world. There is no doubt, too, that Islamic constitutional law offers a very different paradigm from that proposed by modern international law. It is a paradigm, furthermore, which has been implicitly replicated in modern, secular ideas of political organisation, in the Arab world at least, where Arab nationalism has played a major role.

As far as the Islamic state and sovereignty are concerned, very different factors operated from those involved in international law: the basis of the Islamic state was ideological, not political, territorial or ethnical and the primary purpose of government was to defend and protect the faith, not the state (Lambton 1981: 13).

Indeed, the concept of the state was unitary and communal, in that ideally the whole of the Islamic world should form a single political unit, the *umma*, in which society was so ordered by its political institutions that the precepts of Islamic law, the *shar'ia* which ensured the proper observance of Islam, could themselves be properly fulfilled. The *umma* was legitimised by being under the control of a *khalifah* – who embodied the delegated temporal authority of the Prophet Muhammad who was also *imam* – and thereby authorised to interpret the *shar'ia*. In other words, he could pronounce on the temporal consequences of a divinely ordained political order. Sovereignty (*siyada*), then, was seen as a divine attribute, not an inherent attribute of a secular political construct or of authority within such a construct (Joffé 1989: 231).

The crucial aspect of this type of constitutional structure was that it was concerned primarily with community and not with territory. Sovereignty was exercised over the community and, by definition therefore, there could be no prior notion of sovereignty over unoccupied territory. Equally, these constitutional principles also applied, even when the ideal unitary structure did not exist in reality. The fact that smaller, discrete political units existed was frequently the practical reality, of course, but these were seen usually as realms of secular authority and power – *sultah* – under the control of a sultan or amir. None the less, they were legitimised by the fact that they also ensured appropriate political structures for the ordering of Islamic society – the community of believers, in short. It is in the unity of *din wa dawla* (religion and state) that such entities derive their legitimacy (Piscatori 1986: 147–9).

As Bernard Lewis points out, 'The juristic principle of Islam was that the headship of the state was elective' (Lewis 1973: 195–7). This continued to be the case, even when secular authority was dominant, and prevented the formulation of an alternative regular and accepted principle of succession. In practice, of course, secular systems – sultanates and petty dynasties – also used reward and punishment to ensure power, rather than an Islamic archetype. It was only with the Mongols that the concept of family control of state sovereignty develops, so that sons can justifiably succeed their fathers. This was perfected by the Ottomans, although the elective caliphate still remained the ideal. The Ottomans were able, through their concept of dynastic succession, to begin to fracture the old tradition of sovereign authority based on communal sanction and, instead, to look towards a territorial basis for their sovereign power (Lewis 1973: 197).

THE ROLE OF NATIONALISM

It is in the process of legitimisation of political authority that the great differences between modern international law and the traditional

3

Islamic constitutional view of the state are most evident. In sociological terms, certainly from a Weberian perspective, much of the conventional definition of the state is equally acceptable from both standpoints. Its crucial feature is that it is an entity equipped with defined territorial boundaries and with administrative structures that provided efficient means to, in Weber's terms, 'monopolise(s) legitimate violence over [the] given territory' (Scruton 1982: 446–7). Although, in Islam, territory is replaced with community, the state's right to use 'legitimate violence' is sanctioned and sanctified by the *shar'ia* itself.

This definition, however, really begs the question for it requires the violence used by the state to be 'legitimate'. In the Islamic world there is no problem, for it is Islam itself which is the source of political legitimacy. In so far as Islam is a moral and ethical structure, the Islamic state also corresponds to another well-established definition of the state: that of Hegel who saw the state as, 'the actuality of the ethical idea' (Scruton 1982: 446–7). In Europe, however, an alternative form of legitimisation had to be developed, particularly after the Peace of Westphalia (1648) which confirmed the destruction of the religious legitimisation of the European state. The process had been begun by the Augsburg Compromise of 1555 which introduced the principle of *Cuius regio, eius religio* into state practice.

In fact, in European political systems, at least since the end of the eighteenth century, the sovereign rights of the state have been considered to be legitimised by the nation which inhabits it, over which it rules and for which it exists. That was, after all, one of the major achievements of the French Revolution, as defined in Article III of the *Declaration of the Rights of Man*, which stated that, 'The Nation is essentially the source of all sovereignty' (Paine 1969: 132). The sovereign state had become the sovereign nation-state, in which states were legitimised by the fact that their individual territories were inhabited by, 'unique and specific communities, identified by a set of unique cultural values and perceiving themselves to be unique – nations, in short' (Joffé 1987a: 21).

This view of the nature of the modern nation-state is obviously essentially European in origin. Nevertheless it has become generalised, largely as a result of the colonial experience. This is particularly the case in the Middle East and North Africa where both the colonial experience and the integration of the region into the international marketplace because of its role as an oil producer have been responsible for the adoption of the nation-state as the modern paradigm, despite the continued role of Islam. The independent post-colonial states of the region

4

have had to accept the general principles of peaceful relations between states enshrined in international law. They have also acquired an atavistic desire to emulate the successful mode of social and political organisation created by their European precursors – the nation-state – whether or not it is the appropriate ideological model.

In practice, this has proved quite acceptable within the region, provided two conditions are met. Firstly, the territorial extent of the nation-state must be definable, either solely in terms of contemporary evidence of the state's exercise of jurisdiction, or by the historical record if, in addition, there is also substantial evidence of the state's historical cohesion and of its existence as a discrete territorial entity. Territorial sovereignty has, in short, replaced communal sovereignty as the primary attribute of state authority in the Middle East and North Africa.

Secondly, there must be a recognisable nation to inhabit the state and thus to legitimise its institutions and its claim to jurisdiction. It is at this level that the major problem has developed within the Islamic world. There is clearly a potential contradiction between the concept of a universal community legitimised by its attachment to a specific set of metaphysical values that have specific political correlates and the concept of the innate legitimacy of a national community which is culturally distinct from other such communities. The contradiction is aptly described by the major theoretician of modern nationalism, Ernest Gellner:

> In brief, nationalism is a theory of political legitimacy, which requires that ethnic boundaries should not cut across political ones, and, in particular, that ethnic boundaries within a given state – a contingency already formally excluded by the principle in its general formulation – should not separate the power-holders from the rest.
>
> (Gellner 1983: 1)

THE NON-EUROPEAN STATE

Should either of these conditions not be fulfilled, problems are likely to arise over the state's legitimacy in the eyes of those over whom it claims jurisdiction and over the respect paid to its claims to territorial sovereignty by other states. Indeed, it is precisely because of such problems that nation-states in the Middle East and North Africa have encountered difficulties in establishing their popular legitimacy in recent years. The result has been that states in the region have been classified under typologies quite separate from that either of the nation-state or the Islamic *umma*. In one recent publication (Luciani 1990), for example, five different categories are suggested, none of which bear any relationship to the modern concept of the nation-state, although two do depend on Islamic legitimisation for their coherence (Harik 1990: 5–6). The other three are all 'defective' state systems (Heiberg 1975: 186–93) in that, lacking any innate mechanism for legitimisation, they have to rely on repression to sustain the elites which control the institutions of the state.

Interestingly enough, however, the repressive nature of the state is itself justified by the need to create a sense of common consensus within the community it controls – the state's use of violence being legitimised by the general acceptance of the ethical concept of the state itself. That general acceptance implies a popular awareness of the political uniqueness of the population concerned, since otherwise it could look towards other forms and instances of political authority. In other words, all these systems aim towards the process of 'nation-building' in order to create the legitimacy inherent inside the nation-state ideal. Yet, to do so in the absence of a defined and recognised 'nation', they often have to lay claim to an Islamic mode of legitimisation, thereby questioning the basis of the very legitimacy they seek. If they do not do this, they have to seek legitimisation through the Khaldunian concept of *asabiya* – group solidarity of the ruling elite and its innate superiority (the justification for its political role) over the rest of the society it controls (Salame 1990: 31–4).

These contradictions have manifested themselves in different ways in North Africa, compared to the Middle East. This is partly because of the different historical evolution of the concept of the state in each subregion. It is also because of the very different histories of state formation in each case. The states of North Africa, after all, developed directly out of the administrative divides created by the French colonial system there. Only Libya experienced a different form of evolution, as a result of United Nations' stewardship up to 1951, which imposed independence on a state structure originally created by Italian occupation. In the Middle East, on the other hand, the territorial extents of the state there arose partly as a result of British colonial and oil interests in the Gulf region and partly from the competition between Britain and France for control of the Levant after the First World War.

THE DEVELOPMENT OF THE STATE IN NORTH AFRICA

The formal creation of independent states in North Africa from the colonial possessions of France, Italy and Spain has been effectively governed by the principles of international law enshrined in the declaration of the United Nations (UN) and the Organisation of African Unity (OAU). However, even though the newly independent states concerned – Libya, Tunisia, Algeria, Morocco and Mauritania – profess to adhere to these principles, the region has been riven by disputes over their territorial limits and the delimitation of their national borders ever since decolonisation began in 1951. These disputes clearly stem from the actual history of the colonial period and the decolonisation process (Joffé 1987a: 24–5). However, they also stem from traditional ideological assumptions over the nature of the state in the region which are based on Islamic precept (Joffé 1990: 230–6).

The decolonisation process

On the face of it, the process of decolonisation and the definition of new, independent political entities is straightforward. If mandated and trust territories are excluded, the process is determined by Article 73 of the United Nations (UN) Charter. Colonial powers are obliged to:

> recognize the principle that the interests of the inhabitants of these territories are paramount, and accept as a sacred trust the obligation ... to develop self-government, to take due account of the political aspirations of the peoples, and to assist them in the progressive development of their free political institutions.
>
> (United Nations Charter; Article 73)

The basic principles of the UN Charter have been further defined by UN Resolution 1514 (XV) (December 1960) which makes it clear that the principles of Article 73 apply to 'national territories' and that 'the integrity of their national territory shall be respected' (Paragraph 4 of UN Resolution 1514). The resolution also states that, 'All peoples have the right to self-determination; by virtue of that right they freely determine their political status and freely pursue their economic, social and cultural development' (Paragraph 2 of UN Resolution 1514). However, it also states that, 'Any attempt aimed at the partial or total disruption of the national unity and territorial integrity of a country is incompatible with the purposes and principles of the Charter of the United Nations' (Paragraph 6 of UN Resolution 1514).

7

Decolonisation, then, involves an act of self-determination by a colonial people. They make a choice between three alternatives: independence, integration into an existing state, or association with an existing state. Only when territorial integrity would be disrupted can this principle be overruled and, conventionally under international law, this only applies to small territories, such as colonial enclaves like those of Ceuta and Melilla on Morocco's Mediterranean coast and currently controlled by Spain (Akehurst 1987: 294–5). By implication, therefore, the act of self-determination also refers to the territory on which the 'people' involved live. By implication, furthermore, the 'people' constitute a 'nation' in the sense of being a distinct ethnic group – a homogeneous community that sees itself to be culturally homogeneous and unique, distinguished by its culture from any other community (Smith 1979: 2–4). The end result of the process should, then, be the creation of a new independent and sovereign nation-state or the voluntary abdication of sovereign control to a specific sovereign authority which already exists.

Territorial sovereignty

In Africa the fundamental question of the territorial extension of such states has been resolved in an eminently pragmatic way. Effectively, African states have agreed to respect the territorial delimitations laid down by colonial powers as the boundaries of the independent states of the continent. The decision was articulated in the Cairo Declaration of the OAU, which was issued on 21 July 1964, although it is implicit in Article III(3) of the OAU's Charter which was promulgated in 1963. The Cairo Declaration unambiguously and 'solemnly declares that all Member States pledge themselves to respect the borders existing on their achievement of national independence' because, 'border problems constitute a grave and permanent factor of dissension' and because, 'the borders of African states, on the day of their independence, constitute a tangible reality' (Brownlie 1979: 11).

In fact, the Cairo Declaration also finds a justification within international legal practice. It is, in essence, based on the legal principle of *uti possidetis juris* (Murty 1978: 169). In effect, the principle requires that successor states accept the international boundaries set by their predecessors. The principle is derived from Roman law and was resurrected in 1812 to resolve border problems which had developed in Latin America as the Spanish empire there began to fall apart. In the end it was abandoned at the start of the twentieth century in favour of a more

8

straightforward legal rule based on physical occupation, mainly because it was found to be too restrictive. This restrictiveness was particularly relevant to cases where adjacent territories which had been administered under different colonial regimes were involved, because the principle could disrupt the unity of pre-colonial states that might otherwise be recreated in the post-colonial period – a view that the Moroccan government supports (Bougaita 1979: 233–74).

Other states, such as Algeria, however, have seized upon the principle of *uti possidetis*, together with that of self-determination, as a way out of the inconveniences of history, for the modern Algerian state did not exist in pre-colonial times in its present form. It was thus hardly surprising that, in November 1963, just after the end of the border war between Morocco and Algeria, the Algerian foreign minister, Abdelaziz Bouteflika reiterated Algeria's belief in 'the unalterable nature of the borders inherited from the colonial system' (Trout 1969: 428). There does, however, still seem to be a problem, for implicit in the Algerian position over decolonisation is the assumption that the colonial territory, as defined by the borders imposed by the colonial power, also defines the 'people' who are to be the subject of self-determination. They are, in effect, those people who live in that territory.

However, the concept of 'people' in the United Nations Charter really reflects the concept of 'nation', since the application of territorial sovereignty by a state is carried out in the name of the legitimising nation. In fact, although it is not expressly stated within the Charter, it is to the creation of a nation-state that the principle of self-determination is primarily directed. In this respect, therefore, the Charter has amplified the normal definition of 'state' in international law.

State and nation in North Africa

The question then arises as to whether or not the state populations of North Africa do constitute 'peoples' (nations) within the meaning of the Charter. The problem is that their traditional linguistic and cultural homogeneity – as Muslims and as Arabo- or Berberophones – extends across the post-colonial state boundaries, although the problems of Arabo- and Berberophones, urban and rural populations and patronage–clientage patterns has also created vast ethnic – in the sense used by Frederik Barthe (1969) – diversity. Furthermore, traditionally political loyalties have often been primarily directed towards entities far smaller than the modern state – to tribe or region and to powerful patrons or patron–client lineages. The simple experience of colonialism, although breaking down many barriers and creating far greater mobility within North African society, did not destroy these basic patterns of political and cultural identity.

The states concerned have reacted in two ways. Firstly, they have all sought a sufficiently strong national consensus within their ethnically heterogeneous and fragmented populations to legitimise their institutions – with varying degrees of success. Furthermore, these attempts at nation-building have had to take place against the background of competing ideologies, such as Islam and Arab nationalism, which are inherently hostile to the process itself (Burrell 1989: 14, 24–6). Secondly, they have also on occasion sought to rectify the extent of their territorial control in order to establish the national legitimacy they seek in terms of territorial sovereignty.

They have also had to come to terms with their colonial legacy which, after all, created the institutions they now use. Quite apart from the physical changes colonialism wrought – in terms of economic structures, demographic change and the alterations of social patterns – North Africans also had to deal at first hand with European political values and assumptions, particularly when national liberation movements began to seek independence. Yet those movements have also had to appeal to a traditionalist support base within the population they wished to liberate. The result has been that they developed an ideology of the state which, at the same time, tried to incorporate Western concepts of the nation-state and indigenous concepts of the Islamic constitutional order (Joffé 1987b: 22). The inherent contradictions in this process have introduced a form of institutionalised instability into the states that were created once Independence was granted.

The consequences today are that North African states face a dual threat. Their assumptions of statehood legitimised by discrete nations are rejected by the growing Islamist movements that exist in all of them. These, to a greater or lesser degree, seek to restore Islamic constitutional purity to the successor states of the colonial empires. Sovereignty is thus an ambiguous concept, with the territorial imperative being contested by the doctrinal communal vision. At the same time, the sense of loyalty to a wider community, be it Islamic, Maghrebi or Arab nationalist, has undermined the utility of precisely delimited boundaries. Yet secular concepts of territorial sovereignty and borders still persist, even if often justified by reference to normative Islamic constitutional concepts. It seems that in this, as in so many other respects, North Africa is condemned to its dual European and Islamic inheritance and to live with the contradictions it generates. North Africa also has to suffer the consequences of its ethnically or linguistically heterogeneous population which has turned out to be a further pressure towards fragmentation of the state or, otherwise, a paradigm for an alternative definition of a national community which governments could never accept.

STATE, SOVEREIGNTY AND NATION IN THE MIDDLE EAST

Many of the same conclusions apply to the Middle East. The major difference is that most modern Middle Eastern states are successor states to the Ottoman empire as well as inheriting colonial political institutions. None the less, there is a profound attachment to the concept of territorial sovereignty, even amongst states such as Iran and Saudi Arabia; states that might have been expected to have shown a preference for Islamic precept.

Iran

Iran is the only state in the region which occupies a territorial extent that is substantially similar to that which it claimed when the colonial period in the region began, at the start of the nineteenth century. The geographical extent of the modern Iranian state dates from the Safavid conquest between 1501 and 1510 (Lapidus 1988: 207) and, 'Modern Iran inherited from the Safavid period (1501–1722) the pattern of state religious, and tribal (uymaq) institutions which would shape its history to the present day' (Lapidus 1988: 571). Even so, in the Gulf region at least, the existence of the autonomous state of the Sheikhdom of Muhammarah up to the reconstitution of Iran by Reza Shah has meant that the nature of sovereignty in the modern Iranian state is very different from that of its predecessor (Joffé 1989: 154).

Interestingly enough, although the political structures of Iran may have been profoundly changed by the Islamic revolution in 1979, there is little doubt that Ayatollah Khomeini accepted that his *velayat-e faqih* – role of the Islamic jurisconsult – would, in practical terms at least, define a successor state to its secular precursor under the Shah. In *Hukusat-i Islami*, he argues, 'Once you have succeeded in overthrowing the tyrannical regime, you will certainly be capable of administering the state and guiding the masses ...The entire system of government and administration, together with the necessary laws, lies ready for you' (Khomeini 1985: 137). The laws referred to comprise the body of *shar'ia* and, as Ayatollah Khomeini himself points out, 'Islamic government is a government of law. In this form of government, sovereignty belongs to God alone and law is His decree and command' (Khomeini 1985: 56). None the less, the 'state' he was concerned with is undoubtedly Iran in terms of its normal geographic definition, as was made clear in his statement proclaiming the formation of the Council of the Islamic Revolution on 12 January 1979 (Khomeini 1985: 246–9).

The Fertile Crescent

Three of the major states in the region – Egypt, Syria and Iraq – have also tried to mobilise the principles of Arab nationalism both as a vehicle for modernisation and to justify their state structures which, in every case, have become the prerogatives of army-based elites. In two of them – Syria and Iraq – these elites themselves are also based on more informal social links to minority groups inside the population at large: the Alawi-s in Syria and the Tikriti-s in Iraq. The result has been that regime survival and the process of nation-building have depended on repression, although in Egypt a limited democratic process has also begun. As a consequence, in the Fertile Crescent at least, although states exist and the concept of defined territorial sovereignty is generally accepted, the legitimising influence of a concomitant national community generally does not (Kienle 1990: 25). Egypt is, perhaps, the exception in this respect, given its long history, homogenised culture and innate sense of national identity.

Attitudes towards territorial sovereignty in Iraq are, perhaps, typical of the other modernising states in the region. The Iraqi state is an unambiguously twentieth-century creation. Formed from the former Ottoman *vilayat-s* of Baghdad, Basra and Mosul, it was first a British mandate after April 1920, then a protected monarchy after August 1921 and eventually an independent state after admission to the League of Nations on 3 October 1932. Even then, many Iraqis considered that its sovereignty continued to be infringed by the restrictive terms of the 1930 treaty between Iraq and Britain, which gave the former mandate power certain specific privileges for a 25-year term.

It was only after the treaty was renegotiated in 1947 and eventually replaced by a Special Agreement that these restrictions were finally resolved. In part, this was made possible because of Britain's decision to join the Baghdad Pact in 1955. The Pact was seen in London as a suitable replacement to ensure those privileges which it wished to preserve (Penrose and Penrose 1978: 45–56, 123). Sovereignty in Iraq, then, has always been couched in terms of territorial sovereignty. This has been particularly demonstrated in its use of the concept of *uti possidetis juris* to justify its claim to Kuwait, despite Ottoman abandonment of that claim in 1913 (Schofield 1991).

Jordan has an anomalous position in this respect, however. Although it was created as a post-First World War mandate, the Hashemite monarchy that was installed there considered itself to have an Islamic mandate as part of the family of the *Sharif* of Mecca. This was transformed during the 1930s and 1940s into a mandate for Arab unity within the Levant – a project which was rivalled by similar but less strongly felt sentiments from the other Hashemite monarchy created after the First World War in Iraq. In the post-Second World War situation, with the creation of the State of Israel in 1948, Jordan had to harbour thousands of Palestinian refugees and also annexed the West Bank.

In consequence, the modern Hashemite monarchy has had to rely on support from the *badu* tribes and has sought to elicit support from its Palestinian majority population by recourse to its claim to a wider Levantine role. As a result, there is not really a national community within Jordanian territory to justify a conventional claim to territorial sovereignty, although East Bank rural Jordan offers a more homogeneous picture. Indeed, King Hussein's decision to abandon Jordanian claims to the West Bank territories in 1988 was an indication of the basic contradiction within the Jordanian state, as it struggles to digest two nations within the unitary structure of the single state.

Saudi Arabia

The modern sovereign political structures of the Arab states of the Gulf region are, in virtually every respect, a testimony to British imperial policy, spurred on by interest in oil and in commercial control. This is particularly true of the small states along the Gulf littoral of the Arabian Peninsula, but, as described above, it is also true, to a greater or lesser extent, of the three major states of the region: Iran, Iraq and Saudi Arabia. Although at least two of these states had a sovereign existence before the British-dominated colonial period in the Gulf region began, the actual form and extent of sovereignty claimed by all of them today clearly shows the consequences of British interest. This, in turn, derived from British concern over access to India; over commerce during the nineteenth century; and over control of oil production during the first half of the twentieth century.

Although Saudi Arabia claims to be an Islamic state controlled by the

13

Wahhabi *imam-s*, it is acutely concerned over territorial sovereignty. Its formal political structure, as a kingdom (*mamlikah*), betrays this concern. In Islamic constitutional theory, monarchical authority – *mulk* (supreme power, sovereignty or right of possession) – is defined as an attribute of Allah, not of temporal power, precisely because it involves the attribution of sovereignty. The term *malik* is, in consequence, normally only used to describe *non*-Islamic or pre-Islamic authority, since an Islamic authority could not lay claim to a divine attribute of this kind (Moss-Helms 1981: 109–10).

It has also been argued, in the context of Saudi Arabia at least, that the adoption of the term *malik* by Ibn Sa'ud in 1932 was an attempt to associate the divine attributes of *mulk* with those of the Wahhabi *imam*-ate rather than an overt acceptance of the secular nature of the new Saudi kingdom. This did, no doubt, play a role. However, it is difficult to ignore other, perhaps more relevant factors. *Mulk* is an absolute quality and does not depend on conditional authority, as is the case with *sultah* – where the conditional social contract explicit in the *khalifah*'s relations with the *umma* is also implicit. In the European context, moreover, monarchy also acquired a quality of divine right and, furthermore, it was hereditary – just as became the case in the Middle East and North Africa. Even though primogeniture was not necessarily the preferred mode of hereditary succession, there is little doubt that the concept of *mulk* was used in Saudi Arabia to justify the succession process being retained within a small, cohesive family unit – rather as Mawerdi argued should be the case with the caliphate and the Quraish (Mawerdi 1982: 8).

It must also be remembered that the term was adopted at a time when relations with Britain as the major regional power were becoming increasingly important to the new kingdom and just in the wake of the appearance of two British-instituted kingdoms – Jordan and Iraq – in the Middle East region. It is, therefore, difficult to reject the Moss Helms argument, particularly when the same decision was made in Morocco at the end of the colonial period in 1958 – apparently for similar reasons. Thus the use of the term by Saudi Arabia, Jordan and Morocco implies a recognition of non-Islamic forms of constitutional definition as well – hence, no doubt, the Saudi pre-occupation with territorial sovereignty.

None the less, only Saudi Arabia of all the states in the region (with the possible exception of Oman) can claim a degree of legitimacy in terms of traditional Islamic constitutional theory, because of the alliance between Muhammad Ibn Sa'ud, then the Amir of Dariya, and the founder of the Wahhabi movement, Shaykh Muhammad Ibn ʿAbd al-Wahhab in 1744. The Wahhabi movement itself had emerged as a coherent doctrine of religious purification derived from Hanbalism only shortly before its alliance with the al-Saʾuds.

The al-Saʾud family itself had only come to prominence two decades earlier as *amir-s* of Dariya, one of the many petty principates of Central Arabia and based in Wadi Hanifa 20 km to the north of Riyadh. The amirate had been founded by Saʾud Ibn Muhammad Ibn Muqrin. The al-Saʾud family had migrated to the Wadi Hanifa from Qatif during the fifteenth century, claiming to be part of the Anaza tribal confederation of Northern Arabia (Moss Helms 1981: 76–7). Control of the urban centre of Riyadh was to be its initial and primary concern, as part of its strategy to take over the Najd.

In fact, despite the considerable influence that the al-Saʾuds were eventually able to exert throughout Arabia as a result of their alliance with the Wahhabi movement, it was only in the twentieth century that they established the essential link with and control over the *badu* population of central and north Arabia which ensured their ultimate territorial expansion. This followed on from a deliberate decision by ʿAbd al-Aziz to encourage the *badu* into a much closer link with Wahhabism and to create the *Ikhwan* movement as the essential component for territorial expansion (Moss Helms 1981: 128).

The failure of the combined al-Saʾud/Wahhabi movement to establish a Peninsula-wide territorial base permanently before the twentieth century was also a consequence of its conquest by other powers. Although it rapidly expanded during the second half of the eighteenth century, eventually controlling the Najd, it was subjugated to Ottoman influence between 1818 and 1824, when Muhammad Ali conquered the region. The al-Saʾud family also lost control of the Najd between 1885 and 1901 to the al-Rashid of the Jabal Shammar, with its ruling members being forced to take refuge in Kuwait (Moss Helms 1981: 77).

Through intermarriage, the control of the Wahhabi movement also became integrated into the al-Saʾud family and its leaders often justified their claim to temporal rights because of their religious status as *imam-s* – in this respect justifying their constitutional position in purely Islamic terms (Joffé 1990: 231–3). However, after the Kingdom of Saudi Arabia was proclaimed in 1932, King ʿAbd al-Aziz was at pains to emphasise his secular rights:

> The Al Saud were therefore forced to validate their rule within
> Central Arabia by virtue of Wahhabi doctrine and to secure inter-

15

national recognition by emphasising the historical rights of their family as Arab and secular rulers and not as Wahhabi Imams. It is notable that in all treaties with the Ottomans and the British, Abd al-Aziz insisted that one of the first provisions should be the recognition of his family's historical rights and his own right to choose a successor. Moreover, he was to claim that ... the territories of Najd and the Badawin world have extended as far north as Aleppo and the river Orontes in north Syria, and included the whole country on the right bank of the Euphrates from there down to Basra on the Persian Gulf ... and that these territories, having been formerly under Al Saud control, were now his by virtue of his hereditary rights.

(Moss Helms 1981: 110)

In fact, the modern Saudi state was forged by conquest between 1901 and 1925, when the Hijaz was finally occupied (Al-Farsy 1986: 42) and, although today it justifies its sovereign status in both religious and territorial terms, it is the territorial component that has come to be the dominant element. It is for this reason, no doubt, that the Saudi government has devoted considerable efforts to asserting its territorial claims against the Arab Gulf states, against Oman and against Yemen in recent years.

The Arab Gulf states

Amongst the Arab states of the Gulf region, perhaps only Oman could claim a substantially continuous sovereign political authority during the past two centuries over much of the territory it controls today. In fact, it was during the latter part of the eighteenth century that the characteristic political division grew up between the secular power of a sultan on the coast and the religious authority of an Ibadi *iman* in the tribal interior (Al-Naqeeb 1990: 44). In short, Oman had, thereby, begun to create a similar political structure to that which existed in Saudi Arabia, except that religious and temporal authority did not coalesce until 1955 when the separate Imamate was suppressed (Pridham 1986: 136).

Although the future Arab states of the Gulf littoral did exist in embryo by the advent of the nineteenth century, they were effectively only coastal trading and pearling posts. Their ruling families stemmed from the great tribal federations of eastern Saudi Arabia – the Aniza in Kuwait, Bahrain and Qatar and the Qawasim in the Trucial Sheikhdoms (later to become the United Arab Emirates). These small coastal settlements formed part of a mercantile complex stretching from Basra towards the Indian and East African coasts (Al-Naqeeb 1990: 8). Their profit-sharing trading system had been integrated into and dominated by Portuguese maritime trade in the sixteenth century. It was eventually suppressed by Britain after 1839 (Al-Naqeeb 1990: 28). It is only after this period that a process of expansion of territorial control around these trading centres begins to emerge which, during the twentieth century under the pressure of Western demands for oil concession arrangements, made the issue of territorial sovereignty acute.

The common factor about the development of all these political entities except Oman was the focus of political power around maritime trade and pearling. Each of them also depended on the abilities of the ruling group either to dominate local tribal structures or to come to terms with the tribes concerned, in order to ensure their own political survival. Although maritime activities were dominant concerns for the coastal settled populations who formed the majority, the populations of the hinterland, whether *hadar* (settled) or *badu* (nomadic within their own *dirah* – nomadic pasture zone – and usually around 10 per cent of the total) also played a crucial role. Although local rulers controlled their urban coastal populations through a paternalistic system of absolute authority combined with consultation through a *majlis*, the nomadic tribes had to be cajoled into submission, often with large financial subsidies. Failure to maintain this system of tribal coalition would soon destabilise the political structure, as rivals to the ruler would then seek tribal support in their own right (Said Zahlan 1978: 4–7).

This complex system of urban absolutism and rural coalition had little to do with any traditional Islamic constitutional system. It was simply based on the successful exploitation of the segmentary structures of local tribes and the patronage–clientage systems that were their urban variant. In this respect, political power depended on the exploit-ation of *'asabiya* – the concept of agnatic solidarity which, according to Ibn Khaldun, provides the driving force for the assertion of political domination in tribal society. Yves Lacoste has argued that this also provides the basis for the creation of patronage–clientage systems as precursors of modern-style class-based societies (Lacoste 1966: 156).

In the future Gulf states, however, ties between the rural hinterland and the coastal focus of political power were far more tenuous. Gulf rulers devoted much attention to avoiding the operation of the process of the circulation of tribal elites which Ibn Khaldun argued formed the underlying theme of political power in the Muslim world (Lacoste

1966: 129–31). In this respect, the hinterland could have a crucial effect on the ability of a ruler to maintain his position:

> The foremost measure of a coastal ruler's strength and prestige was his ability to command the tribes of the interior; his rise or decline in coastal politics could usually be measured by his ability to enforce his authority over the tribal chieftains in the area he claimed as his territory. Conversely, the extent of a ruler's territory was governed by the extent to which the tribes roaming the area would support him in time of need.
>
> (Said Zahlan 1978: 6)

In fact, one of the few recognised assertions of sovereignty was a ruler's ability to collect taxes, known as *zakat*, from tribes whose *dirah* ran across the territory he claimed to rule, in addition to the customs dues charged on coastal trade, particularly pearling. Another was the ruler's ability to protect these tribes if they were raided (Said Zahlan 1978: 6). In effect, therefore, the sovereignty Gulf rulers sought to assert was territorial sovereignty, even though it was expressed in communal terms (by the collection of *zakat*) because it did not involve any statement about the religious legitimation of the ruler's authority, merely an assertion of his power.

In this complex picture of fluctuating territorial control and pragmatic assertion of sovereign rights, Britain's role was, first, to act as guarantor of the small Trucial Sheikhdoms and states such as Kuwait, Bahrain and Qatar against external pressure. Later on, this role became more codified, as Britain began to interfere directly into Gulf affairs, particularly over border definition, and eventually provided the military guarantees required to defend the delineated territorial sovereignty claimed by the states concerned. It was this latter process that was stimulated by British anxiety to capture control of Gulf oil. None the less, the British government proved to be reluctant to force the issue and few borders had been properly delimited and demarcated by the time that independence arrived.

The one major dispute that was settled was the Buraimi oasis dispute which lasted for 40 years until it was finally resolved in 1974. However, by that time, there had been regular disputes over the control of villages in the complex, almost flaring into open warfare in 1931 and causing a British military intervention in 1955. Thereafter Britain imposed boundaries between the states involved which were modified by agreement between Saudi Arabia and Abu Dhabi in 1974 (Drysdale and Blake 1985: 90).

In general, however, the Arab Gulf states entered into independence in 1971 (1961 for Kuwait) with territorial disputes still in being but with a sense of territorial sovereignty well developed. It is very unlikely that any of the rulers in the Gulf would subscribe today to the views of the Residency Agent after his review of boundary status in the summer of 1937:

In his report, the Agent said that the rulers had admitted that they had no fixed frontiers with their neighbours, but that they had given him instead details of what they considered their *ihram* (sacred possession, and therefore inviolable). The only ruler who was absolutely sure of the extent of his territory was Sa'id of Dubai. The Sultan of Sharjah, by contrast, was the only one who refused to state which territory he claimed.

(Said Zahlan 1978: 148)

Territorial sovereignty has, instead, become the essential source of legitimacy for statehood in the Gulf, as it has elsewhere in the Middle East and North Africa. The remaining problem facing states in the region is to create effective national communities against the twin pressures of Arab nationalism and Islam.

REFERENCES

Akehurst, M. (1987) *A Modern Introduction to International Law*, London: Unwin Hyman.

Barthe, F. (1969) *Ethnic Groups and Boundaries*, Boston: Harvard.

Bougaita, B. (1979) *Les frontières mèridionales de l'Algèrie*, These Illieme Cycle, Paris-I.

Brownlie, I. (1979) *African Boundaries: A Legal and Diplomatic Encyclopaedia*, London: Hurst and RIIA.

Burrell, M. (ed.) (1989) *Islamic Fundamentalism*, London: Royal Asiatic Society.

Drysdale, A. and Blake, G. (1985) *The Middle East and North Africa: A Political Geography*, Oxford: Oxford University Press.

Al-Farsy, F. (1986) *Saudi Arabia: A Case Study in Development*, London: KPI.

Gellner, E. (1983) *Nations and Nationalism*, Oxford: Blackwell.

Harik, I. (1990) 'The origins of the Arab state system', in G. Luciani (ed.), *The Arab State*, London: Routledge.

Heiberg, M. (1975) 'Insiders/outsiders: Basque nationalism', *Arch. Europ. Social.*, VI, 4.

Joffé, E.G.H. (1987a) 'The International Court of Justice and the Western Sahara dispute', in R. Lawless and L. Moynahan, *War and Refugees: The Western Sahara Conflict*, London: Pinter.

—— (1987b) 'Frontiers in North Africa', in G. Blake and R. Schofield,

Boundaries and State Territory in the Middle East and North Africa, Wisbech: Menas Press.

—— (1989) 'International law, conflict and stability in the Gulf and the Mediterranean', in C. Thomas and P. Saravenamuttu (eds), *The State and Instability in the South*, London: Macmillan.

—— (1990) 'Concepts of sovereignty and borders in North Africa', *International Boundaries and Boundary Conflict Resolution*, Durham: IBRU.

Khomeini, Imam R. (1985) *Islam and Revolution: Writings and Declarations* (trans. H. Algar), London: KPI.

Kienle, E. (1990) *Ba'th v Ba'th: The Conflict between Syria and Iraq 1968–1989*, London: I.B. Tauris.

Lacoste, Y. (1966) *Ibn Khaldoun: naissance de l'histoire passé du tiers monde*, Paris: Maspero.

Lambton, A.K.S. (1981) *State and Government in Medieval Islam: An Introduction to the Study of Islamic Political Theory: The Jurists*, Oxford: Oxford University Press.

Lapidus, I.M. (1988) *A History of Islamic Societies*, Cambridge: Cambridge University Press.

Lewis, B. (1973) *Islam in History; Ideas, Men and Events in the Middle East*, Alcove Press.

Luciani, G. (ed.) (1990) *The Arab State*, London: Routledge.

Mawerdi, A-H. A. (1982) *Les statuts gouvernementaux* (trans. E. Fagnan), Le Sycomore.

Moss Helms, C. (1981) *The Cohesion of Saudi Arabia*, London: Croom Helm.

Murty, T.S. (1978) *Frontiers: A Changing Concept*, New Delhi: Palit & Palit.

Al-Naqeeb, K.H. (1990) *Society and State in the Gulf and the Arab Peninsula*, London: Routledge.

Paine, T. (1969) *The Rights of Man*, Harmondsworth: Penguin.

Penrose, E. and Penrose, E.F. (1978) *Iraq: International Relations and National Development*, London: Benn.

Piscatori, J. (1986) *Islam in a World of Nation-states*, Cambridge: Cambridge University Press/RIIA.

Pridham B. (1986) 'Oman, change or continuity', in I.R. Netton, *Arabia and the Gulf: From Traditional Society to Modern State*, London: Croom Helm.

Said Zahlan, R. (1978) *The Origins of the United Arab Emirates: A Political and Social History of the Trucial States*, Macmillan.

Salame, G. (1990) '"Strong" and "weak" states: a qualified return to the Muqaddimah', in G. Luciano (ed.), *The Arab State*, London: Routledge.

Schofield, R. (1991) *Kuwait and Iraq: Historical Claims and Territorial Disputes*, London: RIIA.

Scruton, R. (ed.) (1982) *Dictionary of Political Thought*, New Haven: Yale University Press.

Smith, A.D. (1979) *Nationalism in the Twentieth Century*, Oxford: Martin Robertson.

Thomas, C. (1985) *New States and Sovereignty*, London: Gower.

Trout, F.E. (1969) *Morocco's Saharan Frontiers*, Geneva: Droz.

Wallace, R.M.M. (1986) *International Law*, London: Sweet & Maxwell.

2

TRANSBOUNDARY INTERACTION AND POLITICAL CONFLICT IN THE CENTRAL MIDDLE EAST

The case of Syria

Alasdair D. Drysdale

INTRODUCTION

Traditionally, political geographers who study boundaries have focused mainly on their evolution, morphological characteristics, and conflict-generating potential. Increasingly, however, some have concerned themselves with the functions boundaries actually perform, particularly as barriers to spatial interaction. One of the most obvious, and important, differences among boundaries is the degree to which they are permeable. What happens at boundaries? What flows across them?

This chapter examines the flow of people and goods across Syria's boundaries to and from contiguous or near-neighbouring states. Its central thesis is that the level of transboundary interaction between states is a good measure of the temperature of their political relations.[1]

Generally, geographers have relied on gravity-based models to predict spatial interaction, explaining it primarily in terms of distance, mass, economic complementarity, and functional specialisation. In Europe and North America such variables may, indeed, accurately predict flows of people and commerce between places. But in the Middle East, the level and direction of interaction is often determined by political factors, and transboundary flows cannot be divorced from the political environment in which they occur.

The Middle East has been a zone of conflict and instability and lags behind most other parts of the world in regional cooperation plans. Since the Second World War, it has seen Arab–Israeli wars in 1948–9,

1956, 1967, 1973, and 1982 (and almost in 1991). Between 1980 and 1988 Iraq and Iran were at war, and during 1990 and 1991 the entire region was convulsed by the crisis triggered by Iraq's invasion of Kuwait. Civil wars have preoccupied Lebanon, Iraq, and Jordan, and internal political upheavals have affected most states in the region. The list of states that have had serious quarrels with neighbours at one time or another is long indeed. These conflicts have been profoundly disruptive to the movement of people and goods. Often, boundaries have been closed altogether – between Algeria and Morocco, Libya and Egypt, and Syria and Iraq, for example – making it difficult or impossible to travel from one part of the region to another. At other times, states have demonstrated their displeasure with neighbours by imposing cumbersome and time-consuming border-crossing formalities and thereby discouraging all but essential travel. Conversely, a *rapprochement* between two states will often be accompanied by a well-intentioned, and invariably short-lived, commitment to make their boundary more permeable (Libyan leader Muammar al-Qadhafi once personally drove a bulldozer through Libya's border-crossing post with Egypt – to President Mubarak's indulgent bemusement – to demonstrate his conviction that there should be no restriction on movement between the two countries).

Syria provides a particularly good setting to examine how political factors shape interaction patterns in the Middle East. First, it is situated at the heart of that part of the Middle East where conflict has been most intense, and its relations with other states in the region have been turbulent (Seale 1988).

Partly because of its location – it borders Israel, Jordan, Lebanon, Iraq, and Turkey – it has been directly or indirectly involved in all of the region's major disputes. Few states in the region have doors that open to so many distinct, but interacting, geopolitical realms. In addition, Syria is pivotally located with respect to the region's key trade routes and historically has served as a routeway between Europe and Asia, Africa and Asia, and the Mediterranean and the Persian Gulf. This location has given it immense strategic leverage, which it has exploited with great effectiveness. All of Lebanon's land trade must pass through Syria, and until the early 1980s most of Iraq's oil exports crossed Syria to the Mediterranean. Historically, most of Jordan's trade also passed in transit through Syria to Beirut in Lebanon before it developed a port at Aqaba to ease its quasi-landlocked condition. Between 1967 and 1975, when the Suez Canal was closed, Jordan was forced to rely on trans-Syrian trade routes even though Aqaba was open.

INTERACTION WITH LEBANON

The relationship between Syria and Lebanon is one of the most complex in the region (Weinberger 1986).

On the one hand, many Syrians and some Lebanese still do not fully accept the post-First World War division of the Levant by the British and French, and view the distinction between the two countries as artificial. Syria still has no formal diplomatic relations with Lebanon and has lingering irredentist claims over the country. President Asad once remarked that Lebanese and Syrians 'are one single people, one single nation ... The feeling of kinship runs deeper than it does between states in the United States' (*New York Times*, 4 December 1983: A4).

In fact, ethnically, culturally, and linguistically, there is little difference between Syrians and Lebanese, whose high level of interaction has been helped by close geographic proximity – Damascus is only 75 miles from Beirut. On the other hand, many Lebanese, particularly Maronite Christians, deeply resent Syrian interference in Lebanon's political life, which has almost completely eroded the country's independence. Syrian forces have been actively involved in Lebanon since their intervention in 1976 during the civil war.

Fluctuations in the number of Lebanese arriving in Syria each year since 1958 can be explained at least partly by political factors (Figure 2.1).[2]

The sharp increase in arrivals in 1959 probably reflects the continuing influx of people fleeing Lebanon's civil war in 1958. The drop in 1963 coincided with the advent to power of a radical Ba'thist regime in Damascus following a bloody military take-over, after which an austere Syria turned sharply leftward and inward while freewheeling Lebanon

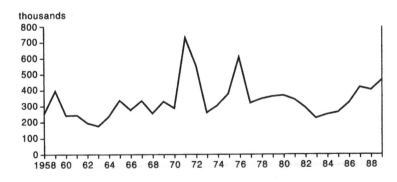

Figure 2.1 Lebanese arrivals in Syria (in thousands)

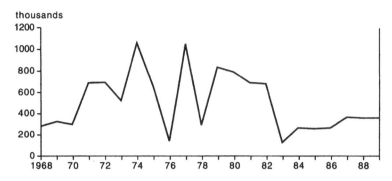

Figure 2.2 Syrian departures to Lebanon (in thousands)

continued to espouse *laissez-faire* capitalism. Similarly, troughs in 1966 and 1970 may have been associated with two more Syrian *coups d'état* and the atmosphere of uncertainty that accompanied them. The decline in arrivals in 1968 might reflect the continuing aftermath of the 1967 Arab–Israeli war. However, one resists the temptation to attribute everything to politics to prove a point: almost constant turmoil in the Levant ensures that one can always find some political event to explain fluctuations in the flow of people across boundaries.

The evidence seems far clearer that politics were at work in 1971 and 1976, when there were sharp increases in arrivals from Lebanon. In 1971, the number of arrivals peaked at over 700,000, more than double the previous year. This surge is easily explained. The November 1970 coup that brought Hafiz al-Asad to power in Syria resulted in an abrupt change in the regime's direction: Asad tried to end Syria's isolation in the region, vowed to liberalise the economy and open up the political system, and invited the large number of Syrians who had fled to Lebanon in the mid-1960s to return home. The drop in 1973 and 1974 may reflect the October 1973 Arab–Israeli war, which obviously disrupted travel in the region, but it may also have been a return to more normal pre-1970 levels of interaction. The sharp increase in the number of Lebanese arrivals in 1976 is clearly associated with the civil war in Lebanon, which produced a wave of refugees fleeing the country, while the decline in 1982 and 1983 may reflect the Israeli invasion and occupation of Lebanon. Since 1984, there has been a steady rise in the number of Lebanese arrivals; the absence of sharp fluctuations suggests there may have been some adaptation to Lebanon's chronic strife.

Data on Syrian departures to Lebanon over the same period appear to confirm the role of politics in shaping interaction levels (Figure 2.2).

The sharp drop in the number of Syrians entering Lebanon in 1973 was caused by the Arab–Israeli war. Similarly, the precipitous drop in the number of Syrians travelling to Lebanon in 1976 and 1983 reflects the civil war and the Israeli invasion (although why there was no decline in 1982 – the invasion occurred in June – remains unclear). The two peaks, in 1974 and 1977, occurred after war-related troughs, and may result from deferred travel or, more likely, may simply measure the large number of people returning home as political conditions improved. There seems to be no obvious explanation for the dramatic decline in Syrian departures to Lebanon in 1978, although in March of that year Israel launched a large-scale incursion into the southern part of the country.

Historically, Syria and Lebanon have been closely integrated geographically and economically. Before the development of Syria's own ports at Ladhiqiyah and Tartus in the late 1950s and 1960s, almost all of its trade passed through the Lebanese ports of Beirut and Tripoli, the natural outlets for central and southern Syria. Economic links between the two countries remain strong. Nevertheless, official Syrian–Lebanese trade figures have almost no meaning (although on the surface they also have been a barometer of conflict, with steep declines during the height of the civil war and following the 1982 Israeli invasion). Despite the almost continuous turmoil in Lebanon since 1976 and the collapse of its central government and infrastructure, Syria remains highly dependent on imports smuggled in through its highly porous border with Lebanon with the active complicity of corrupt customs officials, border guards, and army officers. Whereas Syria's inefficient and stagnant economy has been tightly controlled by a nominally socialist government, entrepreneurial Lebanon has always nourished a vigorous private sector – even in the midst of political chaos. Traditionally, Lebanon has been the lung through which Syria breathes, the principal supplier for its enormous 'black' economy. It is a measure of the Syrian regime's economic mismanagement that the Lebanese, despite being battered by years of civil war, have been selling luxury goods and necessities to the Syrians, not the other way around. Several Lebanese towns immediately across the border cater almost entirely to the Syrian market: shops are stocked full of French cheeses and perfumes, German automobile parts and wines, Swiss pharmaceuticals and chocolate, Japanese electronics and appliances, and American cigarettes and clothes. When a government committee established to combat smuggling invited Syrians to register illegally imported televisions and pay a small fine in exchange for immunity from prosecution, some 155,000 illegally imported sets were declared (*Middle East Economic Digest* 1990: 31).

Some of the Syrian regime's leading figures are widely believed to be deeply involved in this unofficial trade. There are no reliable estimates of the total value of goods smuggled into Syria illegally from Lebanon. However, by all accounts the figure is enormous and, more to the point, not reflected at all in official trade data. Government statistics show a significant decline in trade between the two countries between 1981 and 1986. In fact, the opposite was probably true: official trade declined precisely because there was such a large increase in unofficial trade. Since 1987, official trade has grown dramatically primarily because economic liberalisation in Syria has diverted imports from unofficial channels.

INTERACTION WITH JORDAN

Syria's relations with Jordan, its neighbour to the south, have often been strained (Nevo 1986).

Despite their contiguity, the two countries differ in fundamental ways, and have often disagreed sharply about the best way to resolve some of the Middle East's key problems. King Hussein has steered a moderate course for Jordan, maintaining close links with the West and favouring capitalist development, whereas Syria, at least until the recent collapse of communism, relied heavily on the Eastern Bloc countries for military and diplomatic support and built an economy dominated by the public sector. Generally, Jordan identified itself with the moderate camp in the Arab world, whereas Syria traditionally sided with the radical states. The most serious quarrels have been over Arab strategy in the conflict with Israel. One of the Asad regime's biggest fears has been that Jordan would reach a separate peace with Israel, leaving it exposed and isolated. One of Jordan's major goals has been to escape domination by the foremost Arab power in the Levant region. Since the mid-1970s, at least, many in the region have accused the Asad regime of harbouring irredentist designs and of trying to create a Greater Syria that included Lebanon and Jordan (Pipes 1990).

There is certainly much evidence that Asad tried to bring these countries into Syria's orbit through intimidation, pressure, and even direct intervention. Jordan's alliance with Iraq, Syria's nemesis, was at least partly motivated by the need to counterbalance Syrian power. During the Iran–Iraq war between 1980 and 1988, Syria sided with Iran while Jordan backed Iraq. Similarly, the two countries pursued very different policies in the 1990–1 Persian Gulf war, with Jordan reluctant

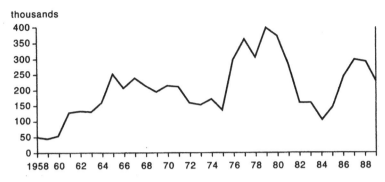

Figure 2.3 Jordanian arrivals in Syria (in thousands)

to desert its former ally. Despite their differences, Syria and Jordan have also been on friendly terms for long periods, particularly in the mid- to late-1970s and since the mid-1980s. Jordan has been reluctant to antagonise Syria, through which some of its trade still passes, and Syria recognises the importance of building a united eastern front against Israel and exerting some influence in Amman.

The ups and downs in Syrian–Jordanian relations are mirrored in the level of cross-boundary interaction, especially after the mid-1970s (Figure 2.3). Between 1965 and 1975, evidence to substantiate the hypothesis that interaction levels reflect the political temperature is mixed. Arrivals from Jordan seem to have been slightly depressed by the 1966 Syrian coup, the escalating conflict within Jordan in the late 1960s between the monarchy and Palestinian guerrillas, and the 1973 Arab–Israeli war. Syria's temporary closure of its border with Jordan between July 1971 and February 1972 to protest the Jordanian army's defeat of the Palestinians obviously reduced overall traffic, although not by nearly as much as one might expect. Monthly data for each of those two years might show sharp fluctuations, with above normal flows immediately before and after the crisis. Whereas the 1973 war appears to have reduced travel, the large number of refugees created by Israel's occupation of the West Bank in the 1967 war may have increased it in that year. Surprisingly, Syria's brief military intervention in northern Jordan in 1970 during its civil war seems to have had no real effect on the flow of people across the border; once again, monthly data might reveal a substantial, but short-lived, impact.

The most striking aspect of Figure 2.3 is the large increase in cross-border traffic after 1975 (interestingly, official Jordanian statistics indicate that the surge began much earlier, in 1973). Between 1975 and

1979, when it peaked, the number of arrivals from Jordan more than doubled from fewer than 150,000 to some 400,000. This spectacular increase can be directly attributed to improved relations between the two countries. After the 1973 Arab–Israeli war and the first US-mediated disengagement agreements between Israel and Egypt and Israel and Syria in 1974, the Asad regime grew increasingly concerned that Egypt, with American encouragement, intended to pursue a separate peace with Israel. To avoid being isolated regionally and to prevent Jordan from following Egypt's lead, Asad sought to mend relations with King Hussein and construct an eastern Arab front (Drysdale and Hinnebusch 1991).

Following their *rapprochement* – Asad's visit to Amman in 1975 was the first by a Syrian head of state in eighteen years – trade barriers were relaxed and a number of joint economic ventures initiated. However, by the end of the decade, relations once again grew increasingly strained as Asad's concerns that Hussein would be drawn into separate negotiations with Israel mounted. The Syrian regime was also enraged when it uncovered evidence that Jordan was offering sanctuary to Syrian Muslim fundamentalists engaged in a bombing and assassination campaign to unseat the Asad regime. In December 1980, it deployed 35,000 soldiers along the border with Jordan and threatened to march in and destroy these bases. The following year, Jordan accused Syria of attempting to assassinate its prime minister. During the early 1980s, Syria sponsored terrorist attacks against Jordanian embassies, airline offices, and other targets, and Jordan responded in kind. Although their border remained open, strict security checks and deliberately dilatory customs and immigration procedures discouraged travel between the two countries. By 1984, crossings from Jordan had dropped to their lowest level since 1960, and were one-quarter of what they had been in 1979.

In September 1985, the countries' two prime ministers met in Saudi Arabia to work out a truce. Two months later, King Hussein conceded that Jordan had backed Syrian Muslim fundamentalists' efforts to overthrow the Syrian regime in the late 1970s. Following his apology, he travelled to Damascus for the first time since 1979 to seal the reconciliation. In turn, Asad visted· Amman in May 1986 for the first time in nine years. The *rapprochement* was accompanied by a relaxation of travel restrictions, which permitted a threefold increase in Jordanian crossings into Syria between 1984 and 1987.

Syrian–Jordanian trade followed almost exactly the same trajectory, trebling in value between 1975 and 1980, plunging between 1980 and

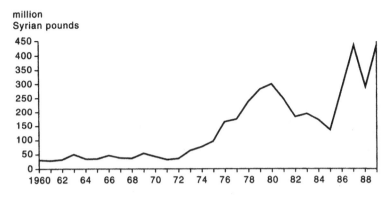

Figure 2.4 Syrian–Jordanian trade (in million £Sy)

1985 as relations soured, and recovering quickly following the 1985 *rapprochement* (Figure 2.4). However, these data are not corrected for inflation. Although there was a substantial increase in the value of trade after the mid-1980s, the volume increase was not commensurate – the rising curve describes Syria's deteriorating economy as much as its improved political ties with Jordan.

INTERACTION WITH IRAQ

No two neighbouring Arab states have had a more tempestuous relationship than Syria and Iraq (Kienle 1990).

The cleavage between the two countries has been one of the most persistent and bitter in the Arab world. Since 1968, both have been ruled by rival wings of the Ba'th Party, whose vituperative internecine disputes have been punctuated only occasionally by ephemeral attempts at reconciliation. Each regime has worked hard to overthrow the other, exploiting every possible weakness. Syria, which depicts itself as the 'beating heart' of Arab nationalism, even went so far as to back non-Arab Iran in its war with Iraq between 1980 and 1988 and to support Western military intervention in the region in 1990–1 to oust Iraq from Kuwait. Since 1982, the border between the two countries has been closed altogether, and Iraq's trans-Syrian oil pipelines to the Mediterranean have been blocked. There have also been disagreements over the allocation of water from the River Euphrates, which flows through Syria to Iraq. The personal acrimony between Hafiz al-Asad and Saddam Hussein has reached the point where a *rapprochement* is inconceivable so long as both are in power.

29

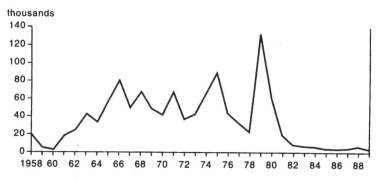

Figure 2.5 Iraqi arrivals in Syria (in thousands)

The relationship between Syria and Iraq during the 1960s was exceedingly complex because of frequent changes of regime in both countries as Ba'thist and non-Ba'thist groups contended for power. For a few months in 1963, the Ba'th Party was in power in both countries, and an unsuccessful attempt was made to unite them with Egypt in a tripartite federation. Between late 1963 and 1968, when the Ba'th was out of power in Baghdad but in power in Damascus, ties between Iraq and Syria were strained. However, this did not appear to dampen travel, which increased during the period (Figure 2.5). The dip in Iraqi arrivals in 1967 was likely due to the Arab–Israeli war. When the Ba'th seized power once again in Baghdad in 1968, any hopes of a *rapprochement* based on the apparent similarity of the two regimes quickly faded. Relations deteriorated sharply as each regime claimed to be the only authentic voice of the Ba'th. The drop in travel between 1968 and 1970 may reflect this tension. However, in 1970 Hafiz al-Asad seized power in Syria, supplanting a more radical faction of the Ba'th, and immediately sought to end Syria's regional isolation and improve relations with Iraq and other Arab neighbours. This entente was also short-lived, as a serious disagreement arose in 1972 over the payment of transit fees to Syria for use of a pipeline to the Mediterranean. Once this dispute was resolved early in 1973, Asad resumed his efforts to build an eastern front and pressed for closer relations with Iraq, which sent forces to assist Syria during the October 1973 Arab–Israeli war. By the mid-1970s, however, relations had once again become frayed. Iraq criticised Syria's acceptance of UN Resolutions 242 and 338, which implied acceptance of Israel, and condemned its intervention in the Lebanese civil war against a leftist-Muslim alliance. A major dispute also arose over water rights following Syria's completion of a dam on the

Euphrates in 1974. Iraq accused its neighbour of withholding water and amassed troops along the border before Saudi Arabian mediation and Syria's release of additional water defused the situation. In 1976, Iraq stopped pumping its oil through Syria, diverting it to a newly completed pipeline through Turkey. As relations deteriorated, the number of Iraqis arriving in Syria fell dramatically, from 90,000 in 1975 to only 20,000 in 1978. Then, in the face of their joint concern over Egypt's moves to make peace with Israel, the two countries once again abruptly set aside their differences and explored the possibility of unification. The almost sevenfold increase in Iraqi arrivals in Syria between 1978 and 1979 was a direct result of this reconciliation. But the *rapprochement* proved just as brief as those before it. Iraq soon accused Syria of subversion, and Syria accused Iraq of having a hand in the wave of bombings and assassinations that rocked the country in the early 1980s. By this time, Asad's anger with Saddam Hussein was so great that he openly sided with Iran in the Gulf war and, as a favour to Tehran, closed Syria's eastern border in 1982 to impede Iraqi oil exports to the Mediterranean. Since then, virtually nothing has crossed the Syrian–Iraqi border. Syria's stance in the 1990–1 Persian Gulf war confirmed the depth of the antagonism between the two regimes.

INTERACTION WITH OTHER MIDDLE EASTERN COUNTRIES

Syrian interaction with other, non-neighbouring Middle Eastern countries has been no less susceptible to political influences. Thus, trade with Egypt fell from roughly £Sy250 million in 1978 to less than £Sy50 million in 1979, the year Anwar Sadat signed a peace treaty with Israel. Between then and 1990 there was no trade, and virtually no other interaction, between the two countries in compliance with an Arab League boycott of Egypt. Direct flights between Cairo and Damascus were only resumed in December 1989, after a twelve-year hiatus, and diplomatic relations were restored shortly afterwards. One can confidently predict that the *rapprochement* between the two countries, marked as it is by all the usual lofty rhetoric about Arab solidarity, will result in a surge of trade.

During the 1980s, one of the most resilient, and in some ways most peculiar, friendships in the Middle East was the one between Syria and Iran. For Syria, the alliance provided a counterweight to Iraq and Egypt and was economically lucrative since Iran rewarded it for its support with free or low cost oil. For its part, Iran gained easy access to

31

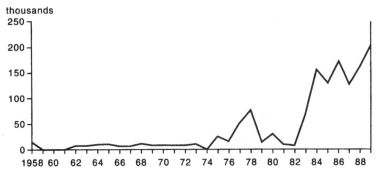

Figure 2.6 Iranian arrivals in Syria (in thousands)

Lebanon's Shiite population and a valuable strategic ally in its war with Iraq. During the 1980s, Iranian tourists flocked to Damascus, virtually taking over many hotels. Indeed, in 1989 over 200,000 Iranians arrived in Syria – slightly fewer than the number of Jordanians, who had to travel only a short distance and faced no language barriers (Figure 2.6). Even in the peak year of 1979, fewer than 140,000 Iraqis arrived in Syria. The massive trade that developed between Syria and Iran in the 1980s was additional proof that interaction between them was grounded in politics.

There is no reason to believe that Syria's regional interaction patterns are anomalous. On the contrary, an examination of flows between most countries in the Middle East would likely show similar sharp fluctuations that could also be attributed to shifts in political relations. The price of conflict in the region has, therefore, been high. If the Arab states, or sub-regional groups of them, are to take seriously their oft-stated goal of advancing regional cooperation, they must somehow ensure that cross-border interaction is insulated from the ups-and-downs in their political relations. Mechanisms must be found to contain political disputes so that they do not cause colateral economic disruption. Furthermore, ensuring that boundaries remain open and permeable encourages interaction and, ultimately, increases functional integration between states, which escalates the costs of conflict to all parties and provides strong incentives to resolve conflicts peacefully and quickly.

NOTES

1 To some extent, political geographers and political scientists have always recognised the political dimensions of spatial interaction, viewing transboundary flows as a measure of political distance between states. See S. Brams, 'Transaction flows in the international system', *American Political Science Review*, 60, 880–98; K. Cox, 'A spatial interaction model for political geography', *East Lakes Geographer* 4, 58–76; K. Deutsch, *Nationalism and Social Communication* (Cambridge, Mass.: The MIT Press, 1966); J.R. MacKay, 'The interactance hypothesis and boundaries in Canada: a preliminary study', *Canadian Geographer* 11, 1–8; R.L. Merritt, 'Locational aspects of political integration', in K. Cox, D.R. Reynolds and S. Rokkan (eds), *Locational Approaches to Power and Conflict* (New York: John Wiley & Sons, 1974); B. Pollins, 'Conflict, cooperation and commerce: the effect of international political interactions on bilateral trade flows', *American Journal of Political Science* 33, 737–61; ibid., 'Does trade still follow the flag?', *American Political Science Review* 83, 465–80; E.W. Soja, 'Communications and territorial integration in East Africa: an introduction to transaction flow analysis', *East Lakes Geographer* 4, 39–57.

2 All data are compiled from the Syrian Arab Republic's annual *Statistical Abstract* (Damascus, 1958–90). Data on arrivals were used because data on departures were available for a shorter time period.

REFERENCES

Brams, S. (1966) 'Transaction flows in the international system', *American Political Science Review* 60, 880–98.

Cox, K. (1968) 'A spatial interaction model for political geography', *East Lakes Geographer* 4, 58–76.

Deutsch, K. (1966) *Nationalism and Social Communication*, Cambridge, Mass.: The MIT Press.

Drysdale, A. and Hinnebusch, R. (1991) *Syria and the Middle East Peace Process*, New York: Council on Foreign Relations.

Kienle, E. (1990) *Ba'th v Ba'th: The Conflict between Syria and Iraq, 1968–1989*, London: I.B.Tauris.

MacKay, J.R. (1958) 'The interactance hypothesis and boundaries in Canada: a preliminary study', *Canadian Geographer* 11, 1–8.

Merritt, R.L. (1974) 'Locational aspects of political integration', in K. Cox, D.R. Reynolds and S. Rokkan (eds), *Locational Approaches to Power and Conflict*, New York: John Wiley & Sons.

Middle East Economic Digest, 16 March 1990, p. 31.

Nevo, J. (1986) 'Syria and Jordan: the politics of subversion', in M. Ma'oz and A. Yaniv (eds), *Syria under Asad: Domestic Constraints and Regional Risks*, New York: St. Martin's Press.

New York Times, 4 December 1983, p. A4.

Pipes, D. (1990) *Greater Syria: The History of an Ambition*, New York and Oxford: Oxford University Press.

Pollins, B. (1989a) 'Conflict, cooperation and commerce: the effect of inter-

national political interactions on bilateral trade flows', *American Journal of Political Science* 33, 737–61.
—— (1989b) 'Does trade still follow the flag?', *American Political Science Review* 83, 465–80.
Seale, P. (1988) *Asad: The Struggle for the Middle East*, Los Angeles and Berkeley, Calif.: University of California Press.
Soja, E.W. (1968) 'Communications and territorial integration in East Africa: an introduction to transaction flow analysis', *East Lakes Geographer* 4, 39–57.
Weinberger, N. (1986) *Syrian Intervention in Lebanon*, New York: Oxford University Press.

3

DEMARCATION LINES IN CONTEMPORARY BEIRUT

Michael F. Davie

INTRODUCTION

Beirut is a word now commonly associated with violence, fanaticism, wanton destruction and traumatised civilians. But Beirut is also a city divided by many boundaries, demarcation lines, internal frontiers, some visible others not, some felt subconsciously, others glaringly perceptible, some short-lived, others nearly a century old. Some of the lines are minor, isolating small quarters, urban blocks or even side-streets; others are major divisions, familiar to newspaper readers or television watchers the world over. The various territories thus demarcated add up to a very fragmented city, struggling to function as an urban entity, its population trying to cope or come to terms with the many obstacles limiting its freedom of movement.

The aim of this chapter is to describe the various demarcation lines in Beirut. The period examined will be 1975–91, although the explanation of the main demarcation line through this city will require going back to the nineteenth century. A very brief model of what constitutes a demarcation line in Beirut will conclude this study.

THE HAZY NINETEENTH-CENTURY BOUNDARY

Unquestionably the most spectacular and well-known Lebanese or Beiruti demarcation line is the one between 'East' Beirut and 'West' Beirut. It is a relatively recent phenomenon; known since 1976 as the 'Green Line'[1] by foreign journalists, it meanders through the heart of what was once the city centre to the suburbs, following in part the 'rue de Damas', the main road leading from the port to Damascus, in Syria. This military demarcation line has existed since the first weeks of the civil war, in 1975, right up to November 1990. It has conditioned all

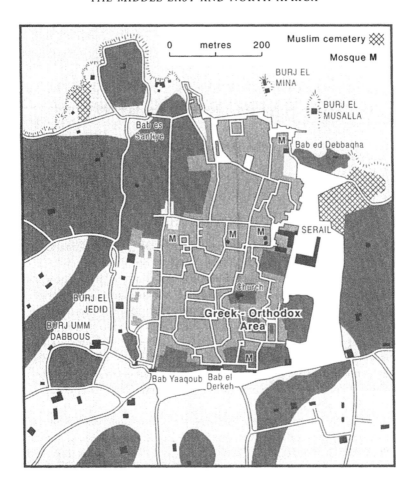

Figure 3.1 Early nineteenth-century Beirut with location of the Greek
Orthodox quarter
Source: From M.F. Davie, 1984.

aspects of life in the city; it has divided populations and given the excuse
for massacres, deportations and destruction; it was the main cause for
the disappearance of multi-confessional quarters and their replacement
by homogeneous ones.

However, a closer examination has shown that the line between the
two parts of Beirut is much older, although it never took on the
hermetic and military aspect of the 1970s and 1980s. Its origin goes

back to the nineteenth century, when Beirut was still a walled city of secondary importance in the Levant. Its population was mainly Sunni Muslim, but with the Christian Greek-Orthodox[2] community constituting around 25 per cent of the total number of inhabitants. These Christians, mainly merchants and craftsmen, lived and worked in a well-defined quarter, to the south-east of the city (Davie 1984, 1987), where their main church and archbishopric was located. Their quarter was quite well placed, in close proximity to the two main gates of the city, the eastern one leading to Tripoli to the north, along the coastal road, and the southern one leading to Sidon and Acre (Figure 3.1), an ideal situation for the local merchants. As required by the Ottoman *millet* system, a religious and secular hierarchy organised all aspects of their life in their quarter. The first 'boundary' was thus between 'Christian' and 'Muslim' quarters. However, one must add the physical limits of the city, the walls, separating the urban from the non-urban. The *souks*, being distinct from residential areas, created other forms of spatial segregation, and thus boundaries; 'rich' areas could be separated from 'poor' ones, the *sérail* quarter from the military barracks, the Turkish quarter from the Arab one, etc. It must be stressed that the 'boundaries' (or spatial limits) were quite permeable and transparent, and that movement was possible from one area to another. The small size of the walled city, the importance of the *souks* and the role of the port guaranteed freedom of movement through economic necessity. Through the *souks*, the city functioned as a complex urban entity, each sector depending on the other, independent of religion. Some sectors would be 'reserved' to a certain community, but business knew no segregation.

The first extension of this area towards the surrounding countryside took place during the 1830s, during Muhammad ʿAli's benign administration. The Christians ventured towards land near their original centre, i.e. to the south-east of the walls; the movement was, however, very slow for lack of a population surplus. This came in 1860 with the influx of Greek-Orthodox from Damascus and Wadi al-Taym,[3] fleeing massacres and destruction during sectarian riots. The Damascene populations fled their own Christian quarter (Schatkowski-Schilcher 1985) and settled in Beirut, then under the protection of French troops; they naturally concentrated around the pre-existing Orthodox nucleus, bought land near to it, and thus developed new 'Christian' quarters (Gemmayzé, Rmayl, Krawiya), to the east and south-east of the old walled city. The Muslim population, without, however, a massive influx of new populations, also increased, and it too started to occupy land on the other side of the walls, extending towards the south-west (Basta,

Zoqaq al-Blat) and, to a lesser extent, to the west, blocked by a topographical spur and Ottoman barracks (Davie 1992).

Two new religiously distinct areas were thus slowly being put into place; the eastern side being mainly Orthodox, the western side being Sunni. However, the eastern side also had a Maronite and Greek-Catholic minority; small nuclei (a few houses, sometimes grouped around a church, or a small sector more or less united by extended family links) would have a different confessional composition in respect to the surrounding environment. To the west of the walled city, it also had sub-areas; Druze or even Christian nuclei (in Zoqaq al-Blat, Qantari, etc.) would be scattered in the predominantly Sunni areas. Limited mixed quarters also existed along the newly built road to Damascus; there is no evidence that any hermetic 'boundary' separated the two communities. Slight differences in wealth, visible in the architecture of newly built houses, different traditions and feasts were the only 'boundaries' separating a basically similar Arab society. Mental boundaries no doubt influenced movement within the city, with some Muslims avoiding some Christian areas, and vice versa, as was the case in all mixed Levantine cities.

The second phase which consolidated this situation emerged with the influx of Maronite peasants from Mount Lebanon during the 1870s and up to the First World War; they trickled down from the mountain in growing numbers, fleeing poverty, over-population and land-ownership problems, hoping to find jobs in town – Beirut had just completed its new port and railway and business was booming (Tarazi-Fawaz 1983). This population settled in the outskirts of the Greek-Orthodox areas, near the port or close to the bridges and administrative limits of the city (Davie 1992). Beirut was Ottoman territory, and its inhabitants subjected to the Empire's laws and military service, while the Moutessarnfate (Autonomous Province) of Mount Lebanon exempted its citizens of such obligations; the Maronite population of Beirut (and to some measure the Greek-Orthodox from Mount Lebanon's mixed districts) chose to combine the advantages of both systems, staying close to the relative security of the Mountain and its different fiscal organisation, while enjoying the prosperity of the city. The net result was a spreading of the eastern side of the city, with a purely Christian population, while the western side extended towards Ras Beirut (with a mixed population: Sunni, Druze and Christian, at Qantari, Ain al-Mrayssé, Zaytouné, Moussaytbeh), or to the south (with a mainly Sunni population, towards Basta Tahta and Basta Fawqa).

The western part of the peninsula is interesting because of the fact that there were several purely Christian quarters and, in one case, a village evolving into a suburb; Moussaytbeh-Mazra'a (Greek-Orthodox) slowly meeting the extending Basta quarter. As for the area around the Syrian Protestant College (later the American University of Beirut), it was, from the very beginning, Orthodox and Sunni, later more generally Christian and Muslim. Christian institutions dotted this western part of Beirut peninsula: European missionary schools, churches, European consulates, even cemeteries. To the east of the city, however, the area was entirely Christian – mainly Greek-Orthodox, with the Maronite quarters on its fringes, notwithstanding some very isolated Muslim sectors (Ras al-Nab'a, Baydoun) on the edges of Achrafiyyé, along the 'rue de Damas'.

THE EAST–WEST DIVISION AND THE DEVELOPMENT OF MIXED AREAS

During the first years of the French Mandate (the 1920s), there was a massive inflow of Armenian refugees (Longrigg 1958), survivors of the Young Turks' massacres, who settled in the eastern part of the city, on land still unoccupied: the Karantina camp near the port, then later in Burj-Hammoud or Karm al-Zaytoun (Kereughlian 1970), further adding to the apparent dichotomy of Beirut. On the western side there was a constant slow influx of Shiites and Sunnis from the countryside or from other countries of the Middle East: the Kurds (Bourghey 1970). The city nearly doubled in size during the period leading up to the 1940s. The city centre retained its commercial function as the *souk* of the city, even though it had been completely remodelled by French urban planners (Buheiry, n.d.). All communities met, as was the case previously in the now-demolished *souks*, on this 'neutral' territory, completely free of any religious spatial differences.

In 1948, there was another influx of populations, this time Palestinian, especially after the Deir Yassine massacre. Once again, religious affiliation seems to have conditioned spatial preferences: the Palestinians, mostly Sunni, settled in camps to the south of the city, on Sunni-owned land (Sabra, Chatila; Tall az-Za'atar being a case apart). The fewer Christian Palestinians settled around churches (Mar Elias Btina), or further north on the coast of Mount Lebanon, at Dbayyé (Gedeon 1974). The main result was, of course, an increase in the Sunni population in the western part of Beirut. However, many well-to-do Palestinians had managed to leave their country in acceptable conditions (Smith 1986), and were ready to invest what capital they had exported in new businesses and commercial activity in Beirut. Together with Lebanese financiers and businessmen they were partially instrumental in the creation of Hamra (Kongstad and Khalaf 1974), the new business district on the western side of the old city centre. Beirut had acquired extraordinary advantages out of the 1948 war: Haifa port was closed to Arab shipping, while Beirut's was the best-equipped of the Levant; banks thrived in the liberal economic status the country had given itself; newly discovered oil in the Persian Gulf and in Saudi Arabia had created revenues that only Lebanese banking know-how could manage in the Arab countries; local man-power was skilled and its elite well trained by the two main universities (American University of Beirut and the Jesuit Université Saint-Joseph); the newly built airport was cosmopolitan and 'modern', with none of the restrictions found in other Arab countries. Most of these advantages were concentrated in the general area around Hamra and its extensions towards Ras Beirut and Ramlet al-Bayda. These areas were confessionally mixed, or, to be more precise, small neighbourhoods would be religiously homogeneous in contact with one or more other small areas, creating a patchwork of sectors. Thus Hamra had a small Sunni quarter around the old mosque, itself surrounded by apartment blocks occupied by families of all religions, not far from the Latin or Greek-Catholic church. A Protestant church was less than a hundred metres from the Koreitem mosque, in a mainly Sunni area.

During the 1950s and 1960s, the country received other populations: Christians of Lebanese or Syrian origin from Egypt, fleeing nationalisations and confiscation of property under the new socialist regime of Gamal 'Abd al-Nasser; Syrian Christians, landowners, important merchants or industrialists fleeing Syria because of similar changes in the political regime, flocked to Beirut. These populations, predictably, settled in the 'Christian' areas, now spreading to the first slopes of Mount Lebanon (Ghorra-Gobin 1983, Ghorra 1975), sometimes organised according to Anglo-Saxon models. Whole quarters were 'colonised' by Syrians from Aleppo or Damascus (Badaro, Suiufi). The Muslim areas also spread, through internal demographic growth (including that of the Palestinian camps) and influxes of rural population.

The 'boundaries' between the two areas in the city centre remained the same, more or less along the 'rue de Damas'; towards the suburbs, the various villages and urban fringes tended to meet, to collide along a line, but never to mix. Each area was confessionally different, each one retained its political or feudal loyalties, each side characterised by a church or mosque. Very few inter-faith marriages took place; each side tended to ignore the other, in the context, however, of reciprocal courtesy.

The first civil war (1958) saw the first real demarcation line dividing the city: it followed more or less the 'rue de Damas', confirming that the two parts of Beirut had opted for two different political choices. The 'Christians' to the east had sided with the Maronite nationalistic political line, while the 'Muslims' followed the Sunni pan-Arab line, with of course many individual exceptions. Barricades were constructed across main roads to block enemy infiltrations from 'the other side'; skirmishes took place between the opposing confessional militias (Qubain 1961). The demarcation line lasted until the end of hostilities and the election of a new president who promptly declared 'no winners and no losers' in an effort towards national reconciliation and unity.

The next period (1960s–1975) saw Beirut spread rapidly with the regular influx of Lebanese rural populations (from Mount Lebanon, the Chouf, the Beqaa and South Lebanon); of Palestinians (after the 1967 war and the battles in Jordan ('Black September') in 1970); of Kurds and Assyrians from Turkey, Iraq or Syria; of Syrian and Egyptian Christians. Europeans and Americans also resided in growing numbers, further developing the Hamra-Ras Beirut sector. Other mixed areas also spread, thanks to the rising middle class. As was the case previously, the settlement patterns more or less followed religious preferences, especially among the lower classes, and homogeneous areas developed: Furn al-Chebbak and Aïn al-Rummané tended to become Maronite; Mazra'a and Tariq-Jdidé were Sunni; the Kurds slowly took over the decayed Jewish quarter at Wadi Abou-Jamil; Burk al-Barajiné and Tall az-Za'atar were Palestinian, each with an ever-spreading belt of Shiite refugees; homogeneous areas were sometimes the result of affinities (Badaro and the Syrian Christians).

However, the main change came with the influx of Shiites moving up to Beirut, fleeing Israeli attacks or reprisals, Palestinian or Lebanese left-wing political or military activity, feudal landlords and general poverty. They settled in Shiite villages or nuclei to the south of the city (Chiyah, Burj al-Barajné, Lailaké). The other mainly 'Christian' villages (mainly Maronite: Haret al-Hraik), now suburbs of Beirut, found themselves overwhelmed by the number of new arrivals, and some preferred to live in 'Christian' areas, further north or east: Hazmiyé, Hadath, Jdaidé, Baouchriyé.

The state apparatus also migrated, leaving the city centre for the outer periphery (Beyhum 1991). The new presidential palace was built at Ba'abda (near the historical capital of the Moutessarnfate), as was the new Ministry of Defence and a television station; the area was also chosen by some diplomats for their residences and by upper-class schools. It must be noted, in retrospect, that the sector was Christian.

BEIRUT AND THE CONFESSIONAL DIVIDE OF THE SUBURBS

The spreading of these villages or suburbs, whether 'Christian' or 'Muslim', led to their meeting: the periphery of the Palestinian camp of Tall az-Za'atar was Shiite, and it touched the spreading 'Christian' village of Dekouané. This inevitably led to tensions along the interfaces of each territory, especially as the suburbs were relatively poor (rural migrant population), badly or not at all integrated into the city's economy, still sociologically very attached to their places of origin, and still very 'feudal' in their political attitudes and opinions (Seurat 1985, Nasr 1985), or, on the contrary, influenced by Palestinian revolutionary or nationalistic ideologies.

It was in these classes that the militias developed and recruited; it was along the contact-lines between the various quarters that the first frictions and incidents took place, paving the way for the all-out civil war. It must be noted that these incidents took place in the suburbs, and not in Beirut itself, confirming that the city (or at least its educated dynamic middle class) was slowly shedding its religious bi-polarism: the city was far more occupied with continuing its prosperity (the hike in oil prices during the 1970s, a general easing of tension in the Middle East after 1973, the consolidation of the wealth of the oil-rich states, influx of capital and investments) than with coping with rising dissent and religious or political intolerance on its periphery. Thus, the war was between the suburbs and their respective populations, and not between the still-'neutral' 'East' and 'West' Beirut city centre.

Briefly, then, Beirut in the early 1970s was already divided into two sectors, each one more or less religiously homogeneous: the 'Christian' side and the 'Muslim' side. However, this classification is far too simplistic and needs once again moderating. In the 'Christian' sector, many different rites could be found: Maronite, Greek-Orthodox, Greek-Catholic, Latin, Protestant (with all the different sects), Copts, Assyrians, etc. They would sometimes be concentrated in specific areas, for historical reasons: Maronites in the suburbs and outskirts of eastern Beirut (Jdaidé, Sinn al-Fil); the Greek-Orthodox and the Greek-Catholics would be concentrated nearer to the city centre, or in completely new areas when the Greek-Catholics came from Syria; in 'Muslim' Beirut, to the west of the 'rue de Damas', areas would be specifically Sunni (Basta Fawqa and Basta Tahta, Zeidaniyé) or Druze (Karakol Druz, small areas in Ras Beirut), or Shiite (parts of Chiyah, Amilieh). Each area could be, in turn, differentiated along socio-economic lines: 'rich' (Hamra, Ras Beirut, Verdun, Achrafiyyé) or 'poor' sectors (Basta, Karm al-Zaytoun, Nahr, Sabra, Naba'a). In turn, one could classify them as 'new' (Ramlet al-Bayda, Sioufi, Mar Taqla) or 'old' (Gemmayzé, Ghalghoul, the Port area, Bab Edriss); populated by Europeans or Americans (Hamra, Koraytem, Verdun) or almost exclusively by Palestinians (Sabra, Chatila, Tariq Jdidé), Armenians (Burj-Hammoud), Assyrians (Hay al-Syriân), etc. Lines could also be drawn around ethnic quarters: Armenian (Burj-Hammoud) or Kurdish (Wadi Abou-Jamil). While the eastern side of Beirut was classified as 'Christian', it also had enclaves of Muslims (Naba'a, Za'itriyé, Berajaoui): for the western side it was more difficult to define 'enclaves', as the population was more readily mixed, and some areas could not be classified on a purely religious basis (Sanây'a, Hamra, Manara), as noted previously. Spatial differentiation criteria were numerous and varied; this did not hamper movement between the various areas, and no physical 'boundary' existed.

In pre-war days, then, inter-area limits did exist – just as in any city – with varying levels of 'visibility'. A poor Sunni quarter would be likely to display posters of Gamal 'Abd al-Nasser, and the shop names could be only in Arabic; a poor Maronite quarter could only have shop names in French, and display portraits of one of the Christian leaders; a Kurdish quarter would be characterised by a different language spoken, different food and music, different dress. In the Armenian quarters the inhabitants would display pictures of Mount Arafat or Erevan, speak in Armenian, write the shop names in that language, read their own news-papers, join political parties that were exclusively Armenian. Thus, the number of boundaries in the city was great, and one would cross into many territories as a matter of course. These territories had their internal politics, problems and tensions with those surrounding them; the level of violence was manageable thanks to the za'im system (Johnson 1986), which slowly became progressively unadapted to solving the social problems of an ever-increasing population. The political control of the core of the political and economic system by the periphery, far more religiously coherent, inevitably led to the collapse of the mixed centre and the imposition of hermetic boundaries between the radically different religious territories.

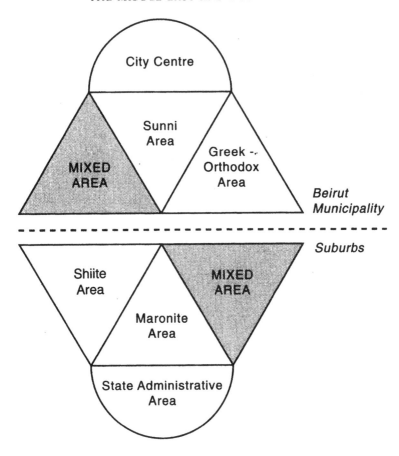

Figure 3.2 Distribution of religious communities in pre-1975 Beirut
Source: From Beyhum 1991.

A schematic diagram (Figure 3.2) illustrates the various areas occupied by the major communities. The city centre, the business area, was an integrated mixture of all the communities: to the south, the Sunni area; to the east, the Greek-Orthodox and to the west (Ras Beirut) another mixed area. The suburbs are different; to the south and south-east were the Maronite areas (Furn al-Chebbak, Ain al-Rummané); to the north-east mixed areas, some poor (Naba'a, Za'itriyyé, Rouaissât) or rich (Rabiyé, Yarzé); to the west, the Shiites (Burj al-Barajné). The southern part was the state administrative centre concentrated around Ba'abda, with the presidential palace, the barracks, the Ministry of Defence. The mixed areas (the city centre, the state administrative area, etc.) occupied roughly half the area of Beirut and its suburbs; furthermore, the various boundaries between the areas were formal and permeable.

THE INTERNAL WAR AND NEW BOUNDARIES

The civil war, when it erupted in 1975, took advantage of all these internal divisions. The 'eastern' side was immediately taken over by the Kataëb militia, massively Maronite and recruiting in the 'Christian' suburbs. On the 'western' side, various militias took over; the Murabitoun and their Palestinian allies were associated with the Sunni population (Johnson 1986) (although not necessarily depending on the traditional leadership); the only 'mixed' militias were left-wing formations which were quickly shunted aside by the increasingly religious groups.

Quite naturally, the confrontation lines were almost exactly those of the major East–West divide, i.e. the 'rue de Damas'. This line must be analysed as being the 'maximum optimum religious extension' for both sides. No major battle took place after 1976 for control of territory 'on the other side' of the line: the defeats of the Christian militias to the west of the city centre in 1975 and early 1976 can be explained by their inability to control non-Christian territory. Military activity settled down to a routine of sniping and occasional shelling from behind well-fortified lines (Davie 1983). All roads on each side were blocked with containers looted from the port, which forbade any forward movement. Each medieval-style 'fortress' was accessible only through 'gates', severely controlled openings through the fortifications (Tabet 1986).

The homogenisation of each side was only relative. In 'West' Beirut, the mixed areas of Hamra, Moussaytbeh, Mazra'a, Ras Beirut and Ramlet al-Bayda continued to function as such, albeit with a declining Christian population. In 'East' Beirut, in contrast, the various massacres (Tall az-Za'atar, 'Black Saturday', Jisr al-Bacha, Naba'a, Karantina) rendered the area almost exclusively 'Christian'.

However, the 'Christian' and 'Muslim' areas were controlled less by a consensus between the population and the militias than by brute force: the various factions often had to silence dissent (expressed through political channels or individually), through censorship, imposed ideology, abolition of all civic freedoms. As the areas were a complex mixture of population or different geographical, social and confessional backgrounds (Figure 3.3), tensions were bound to emerge, and the

Salâm : Name of Quarter
(Sunni) : Confession of Inhabitants

Figure 3.3 Spatial distribution of communities in Moussaytbeh (West Beirut)
Source: From Beyhum 1991.

various sub-groups would evolve into gangs furthering their particular limited interests. Notwithstanding these differences, the militias were able to be at least partially accepted by the population of the territories they controlled, through the use of religiously based slogans. The Kataëb would be 'defending the Christians, a tiny minority struggling for freedom in an ocean of Muslims', while the other side of the 'Green Line' would stress the Crusader-like Western attitudes and alliances of their enemies. The use of religion was a very effective tactic used to consolidate militia power.

The first part of the civil war ended with the deployment of the Syrian-led Arab peace-keeping force. Tension soon erupted between the Syrian contingent and Bachir Gemayyel's Lebanese Forces militia. The Syrian troops retreated to heavily fortified positions (around the city's skyscrapers, or on strategic bridges and crossroads), effectively dividing East Beirut (and especially Achrafiyyé) into smaller controllable areas. Movement between one area and another implied crossing boundaries, often in the cross-fire of the opposing sides. Beirut was again fragmented into areas under the effective control of the militia, the Syrian Army or of the remnants of the Lebanese Army. The 'Green Line' was still present, as action against the Lebanese Forces required opening fire from all sides, including from West Beirut.

New divisions, new boundaries replaced the pre-war ones. The various Christian militias (Kataëb, Harras al-Arz, Tanzim, PNL) each controlled a fragment of 'East' Beirut, setting-up check-points, marking 'their' territory with graffiti (Chakhtoura 1978, Habib 1986) and posters, opening party offices. Skirmishes broke out regularly for the control of contested streets or for 'strategic' crossroads or buildings. Boundaries inside East Beirut, already separated from West Beirut by the 'Green Line', slowly rendered internal movement tedious and, for some, dangerous, as man-hunts were common.

After the evacuation of Beirut by the Syrian troops, and following the elimination of all internal opposition to Bachir Gemayyel, East Beirut was slowly organised into a semi-autonomous area. A militia-run administration, police force, army and political structure were put into place. However, the uniformity of these areas was only superficial: areas could be differentiated by the subtleties of political or ethnic loyalties – the Assyrians retained control of their 'enclave' (Hay al-Siryan) and various portions of the Green Line; local gang leaders were promoted to 'area commanders'; leaders belonging to previous militias remained in command of 'their' men, in more or less the same areas (the PNL in SODECO, for example). East Beirut was thus fragmented, with boundaries between the various territories, although at a different scale and intensity *vis-à-vis* West Beirut. These boundaries were visible by the type and combination of posters of Christian leaders (Camille Chamoun and Bachir Gemayyel would indicate a PNL stronghold; Pierre Gemayyel and his son Bachir, a Kataëb one; Bachir Gemayyel and the party flag, a Lebanese Forces area). Slight changes in uniforms, vehicles or armaments, the newspapers read or available, the relationships between the civilian population and the militiamen, and key-words in conversations all combined to differentiate areas; one could 'feel' a boundary, although it was completely invisible. The previous categories of differ-entiation persisted, of course, between, for example, Greek-Orthodox, Greek-Catholic, 'poor' and 'rich', 'old' and 'new' areas.

What was left of the state's authority was concentrated in a few key areas, around the Ministry of Defence and the presidential palace, on the outskirts of the city. There, the Lebanese Army was present; one would thus leave militia-held territory and suddenly enter areas with Lebanese flags, portraits of the president and soldiers in uniform. In times of tension between the Army and the other armed factions, check-points would be set up, traffic filtered and the different boundaries even more clearly marked.

West Beirut evolved into an even more complex situation. In 1978, a massive Israeli attack created a wave of panic among the Shiite popu-lation of South Lebanon, which surged northwards to Beirut, requi-sitioning all available lodging. In West Beirut, the Shiites took over whole areas, especially the decayed ones or those close to the Green Line. The declining number of Christians slowly filtered over to East Beirut, while the Sunnis tried to preserve some of their own quarters. This Shiite 'invasion' led to fighting for the control of territory: the Shiite Amal militia fighting Palestinians, the Druze PSP, and later the Hizballah, as well as continuing the ongoing war with East Beirut. Inside each territory, the situation was very similar to the one en-countered in East Beirut.

The city was again fragmented into small territories, each with its visible and invisible boundaries. The situation at the end of the 1970s and early 1980s can be synthesised by a diagram (Figure 3.4) with the Shiites entering Sunni territory in Beirut and replacing this community, with the progressive disappearance of the mixed areas in favour of the Shiite influx. To the east of the demarcation line, the Maronites politically replaced the Greek-Orthodox, while the mixed Christian–Muslim areas were flattened. The previously mixed or 'neutral' state administrative area was progressively integrated into the 'Christian' militia area.

Other demarcation lines appeared during this period, during the 1982 invasion of Lebanon and the subsequent siege of Beirut. The demarcation line was between Israeli-held Beirut and 'free' Beirut, between militia-held East Beirut and Israeli-held areas, between West Beirut and East Beirut, these lines cutting across all the previous ones. Even the sea had its line, as there was a naval blockade of the western part of the capital: while ships could come in and out of Jouniyé, in 'Christian' militia-held territory, nothing could enter or leave West

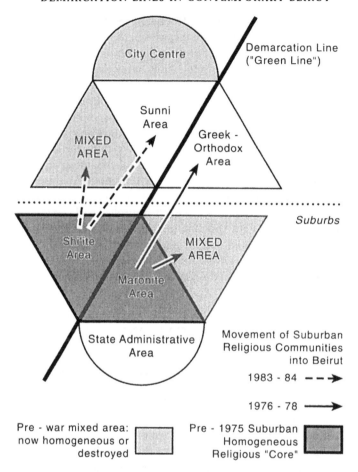

Figure 3.4 Spatial distribution of communities: the situation in 1990
Source: From Beyhum 1991, modified.

Beirut. However, even East Beirut and its suburbs were policed by the Israeli army, with check-points and identification patrols, closed streets and observation posts, creating spatial differences constructed according to military considerations.

Boundaries continued to play an important part in inter-militia relations from 1983 onwards, once Bachir Gemayyel was assassinated. Internal squabbling led to confrontation between rival militia *za'ims* and *qabadays*, each entrenched in his own area, each loyal to particular individuals, while however recognising the overall sovereignty of the militia. In East Beirut, military confrontation led to further fragmentation inside the city and in its suburbs, with each side consolidating its territorial position through alliances with other factions, sometimes even with those of 'the other side'. Lines were drawn according to social, ethnic, economic, morphological and political lines. Newly arrived refugees, deported from the Chouf or from Damour and from other mixed areas, grouped together, maintaining accents, allegiances, attitudes. The new arrivals joined the militias as their only way to get a regularly paid job. Of rural origin, they had no special allegiance to the city and its codes; they had no respect for its inhabitants or its way of life; in return, they were despised by the Beirutis. Socially lower class, they envied the apparent wealth of the Beirutis and their way of life. They settled in areas that soon became self-controlled enclaves (Naba'a) identified by exterior signs (posters, graffiti, accents, flags, etc.), and check-points, observation posts, identification signposts. Similar examples could be given for West Beirut, especially with the emergence of the Hizballah movement in the southern suburbs of the city.

COMPLETE FRAGMENTATION AND SUBSEQUENT 'RE-UNIFICATION'

This period can be described as the paroxysm of the fragmentation process, with completely new boundaries separating hitherto unaffected areas. The situation, at that moment, was extremely complex: on the eastern side of the city, the Christian militias were theoretically unified; practically, they were divided into sub-groups of different geographic origin (Chouf, Bcharré, Damour) or having particular historical evolutions. Some groups traced their origin to the first militia groups of the civil war, in 1975; others were from the same village as one of the leaders; some were unconditional supporters of Bachir Gemayyel's hard line, while others were more conciliatory; part of the militia formed of ex-army soldiers, another by volunteers or schoolchildren. In West Beirut different sub-groups emerged in the Shiite Amal movement, according to whether the militiamen came from South Lebanon or from the Beqaa, whether they had lived in Beirut before 1975 or not, whether they belonged to specific extended families or clans, whether they were close to the Hizballah or a particular *imam* or not. All of these varieties combined to create microscopic territories, each with its own boundary, each with its own code of acceptance.

Even the territory controlled by the state was showing signs of disintegration, with personal loyalties of the officers and troops of the

Lebanese Army taking precedence over loyalty to the country or regime. In time, the Army reacted exactly like the militias, with their specific territory delimited on the map, a situation visible on the ground thanks to numerous indicators. As the Army was in control of the adminis-trative centre (to the south of the city, around Ba'abda), this territory tipped into the hands of a faction, the latest case being General Michel Aoun.

The political and military structure put into place by the militias in East Beirut was challenged by regiments loyal to General Aoun. The first battles, in 1988, created new demarcation lines in the suburbs of the 'Christian' part of the city. Above the usual criteria used to differ-entiate areas, one had now to add that of allegiance (or not) to Aoun's action, acceptance or refusal of his political programme or of his vision of the future of the country. The lines between the areas held by the Lebanese forces and Aoun's army were clear: on one side posters of Geagea, the Lebanese forces leader, slogans stressing the role of the militia in 'Christian' society, their symbol, a stylised cross, sarcastic graffiti about the opposing militia. In February 1989, fighting broke out again with an all-out offensive by General Aoun's troops against Geagea's. The boundary meandered through the city and suburbs where the cease-fire line had stopped the opposing troops: Aoun's army now occupied parts of Lebanese forces territory (Furn al-Chebbak, Ain al-Rummané), while the militia had forced the Army out of Achrafiyyé. During this phase, the 'Green Line' remained unchanged.

The demarcation line between the opposing forces was materialised mainly by main highways or roads, with each opposing faction entrenched on each side. This line was, of course, the consequence of military action on the first days of fighting, and not planned as such; it was, however, very clear on the ground: symmetrically ruined buildings; rubble on the streets; up-ended, mined or booby-trapped containers blocking every side road; sandbags at each window. The passageways into each side's territory were clearly identified by huge flags, posters or religious banners: accents, language and slang would be different (a North Lebanese Maronite accent predominant in Aoun's soldiers; Beiruti, Choufi or Assyrian accent in the militia). Formalities would be different on each side: passports issued here would not be recognised there; car papers, official receipts, etc. would require checking on both sides; different taxes would be levied, different papers read, different radio and television stations listened to. Similarities with the 'Green Line' were striking.

While these new lines were being consolidated (and most 'gates' progressively closed), the old demarcation line between East and West Beirut opened wide for the first time in nearly ten years. The blockade imposed by General Aoun on East Beirut forced the Lebanese forces to make overtures to the opposing side, and especially to the new legal government led by President E. Hraoui. It thus became easy and safe to cross over to 'the other side', purchase the necessary food and petrol, even visit friends. West Beirut was just as fragmented as East Beirut, although it was much quieter, with the end of the internal fighting among the local militias, thanks to a heavy-handed presence of the Syrian Army. This gave a semblance of unity to that part of the city.

In October 1990, with the end of Aoun's rebellion, all of these demarcation lines ceased to have any meaning. In 24 hours, Syrian and Lebanese Army units occupied all of Aoun's territory. Within a couple of months, the militias had been dissolved and disarmed, the Lebanese Army reunited and reorganised, the Constitution modified, and – the most important point – the 'Green Line' dismantled. All roads, including side-roads were opened, de-mined and cleared. No check-points were put into place: tall buildings were demilitarised; all exterior aspects of the militias removed (party offices, slogans, graffiti, flags, etc.). The demarcation line between the positions of the Lebanese forces and Aoun's troops disappeared overnight, with rubble and military positions removed and cleared. The ease and speed with which the boundaries disappeared was proof that many of them were purely differences. The 'Green Line' was perhaps less of a deep Christian–Muslim barrier than the will of militia leaders of both sides to control their own specific territory. The Kataëb, and later the Lebanese forces, used East Beirut to consolidate their own power, to set up structures they directly controlled, to drain revenues from the port or from citizens for the party's coffers. Each militia in West Beirut did exactly the same, with varying degrees of efficiency. Once the militias were weakened and their leaders fallen into disrepute, another force could step in and redefine the territory and modify the boundaries. The latest force in Beirut is that of the state, emerging from an a-territorial position to the uncontested authority governing all previous territories, on both sides of the 'Green Line' and in all sub-territories. Beirut is now an officially 're-united' city, ready for reconstruction; there are no lines that one is forbidden to cross; there is only one authority, the state's, in place of the motley of gangs imposing their own on crumbs of territory.

However, the 'united' Beirut is still very much fragmented: the now 150-year-old East–West divide is still there, and 'East' Beirut is still 'Christian' while 'West' Beirut is 'Muslim', although the term now has very little relevance. Political unification has not erased psychological demarcation lines: many Beirutis, especially those who grew up in the war, still refuse to cross over 'to the other side' (even though growing numbers are exploring the country, after having been cooped-up in ghettos for over 15 years). Perhaps the largest divide to have survived 'unification' has been two different geographies and ways of life, each having evolved along individual ways: East Beirut now has its own circuits (food supply, banks, preferential movements, leisure and entertainment centres), which are quite different from those of West Beirut. An example: the Christian side readily use the beaches to the north of the capital, or move up *en masse* to the 'mountains' (Broummana, Ajaltoun-Reyfoun) for entertainment or residence. West Beirut has no access to the sea because the beach complexes were occupied by the Shiites from 1976–9, and the 'mountains' (Aley, Bhamdoun) were either destroyed or were the front line for over 15 years. The 'Green Line' is still a barrier for the inhabitants of West Beirut looking for a change, while those from East Beirut will not easily cross over for superior restaurants or cultural life. Ethnic, political, linguistic, psychological, cultural *de facto* boundaries are still in place; their importance is now minimised, though not yet erased.

CONCLUSION

The major East–West *de facto* division of Beirut is in no way the only boundary in the city. Ever since the early nineteenth century, spatial differences have existed, implying limits between the various territories, with the criteria used to differentiate them being extremely varied: demography, socio-economic levels, ethnic origin, individual histories, religion and confession, land use, local and regional economy, etc. These variables all interact, but at different levels and geographical scales, with different intensities, and not necessarily simultaneously. Some variables will combine with others at a particular moment in time, but not at another, creating a complex correlation matrix.

In pre-war Beirut, when the city functioned 'normally', movement was possible between all of the various sub-territories, and the boundaries between them were more or less hazy, if not invisible. They had no real 'meaning', as the main preoccupation was keeping the system and the economy in working order. Spatial segregation along one important variable (such as religion) would have unacceptably hindered the economy, which was organised from the city centre, by necessity a mixed 'neutral', multi-confessional territory, where all the variables

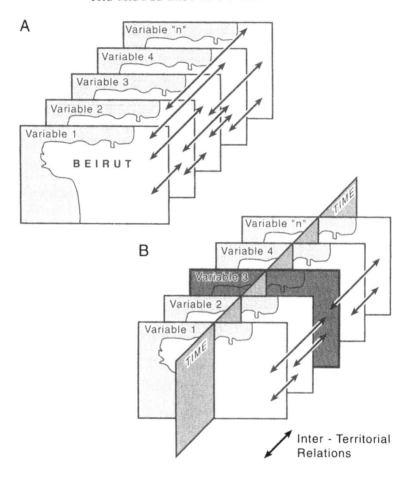

Figure 3.5 Inter-territorial relations before 1975 (A), and during 1975–90 (B)
Source: From Beyhum 1991.

could interact freely, if not cancel each other. The line between 'East Beirut' and 'West Beirut', identified through historical, demographical and religious variables, thus had no impact on the city.

In times of crisis, however, one of the variables could become dominant, completely smothering all the others in importance. In the case of contemporary Beiruti history, this variable was the artificial highlighting of the religious factor, putting into place a spatially segregated 'Christian'–'Muslim' dichotomy. The consequence would be that movement from one territorially expressed variable to another would be hindered, if not rendered impossible. In post-1975 Beirut, the religious variable blocked all movement between the two sectors of the city, and the boundary (the 'Green Line') became hermetic.

It must be noted here that the instigators of this highlighting were not Beirutis but the inhabitants of the city's periphery, often first- or second-generation semi-urban Maronites or Shiites, badly integrated in the city's economy mainly dominated by upper-class Sunni and Greek-Orthodox businessmen, merchants and bankers. The suburbs, located in concentric circles around the city, were the product of mass influxes of rural populations at different periods. Neither purely rural nor purely urban, these countrysides were in a period of social crisis. The political allegiances of the suburban populations were of a different order from those of the city's; less cosmopolitan, less used to the rich mixture of the spatial variables of the city, these suburban populations entrenched themselves in homogeneous sub-territories which were a solution to their political problems and contradictions. They projected this territorial homogeneity onto the city, the centre of financial and social power; instead of slowly becoming part of the system, they chose to overthrow the city and impose their own territorial vision of the world, i.e. homogeneous control, through the religious variable. Beirut city toppled from being a mixed area to two distinct ones, separated by a more-or-less hermetic boundary, the 'Green Line'.

Globally, then, the hierarchical sorting of these variables created territories (and boundaries) which impeded movement according to the criteria chosen (Figure 3.5). The fragmentation of the city, while consolidating local militia power, destroyed the very reason for the existence of the urban system, i.e. the exchange of goods and ideas. In each of the new sub-territories, the internal variables inter-played, creating even smaller territories, each with its own *de facto* limits: conflicts along these new intra-territorial boundaries were common, further fragmenting the Beirut conurbation. The gangs who controlled the sub-areas, well-protected by medieval-style barricades, were unable to make their respective areas function or develop; each area struggled to survive with its local leaders, through ransom, extortion and organised looting. The militias, the war-lords' armies, would then fight for the control of larger areas once they realised the futility and bankruptcy of their respective ghettos. The rallying-call which justified the very existence of these tiny territories (and by extension the boundary between 'us' and 'them') would slowly cease to have any meaning, and the fragmentation of Beirut and its suburbs would continue through inertia; restrictions to free movement between territories would persist, though futile. This complete fragmentation of the city and its peripheral areas completely paralysed the city and, by extension, the country.

Time, then, was a major variable, and none of the territories created during the war lasted for more than a few years. Thus, while the 'Green Line' was a major obstacle, created according to the 'religious' variable, other lines were put into place according to other more important considerations; the 'East–West' dichotomy lost its *raison d'être*, as was amply proven by the Aoun–Geagea conflict (1989–90). At this point, an outside force (either foreign or non-militia) could step in, minimise the separating variable (in this case religion), highlight those variables which 'unite' the Lebanese and the city would be 'transparent' again, functioning more or less 'normally' with all its many other variables. Inter-territory movement would be possible again.

Boundaries in contemporary Beirut are thus less the consequence of historical or geographical determinism than of the political use of one of the numerous spatial variables of the city.

NOTES

1 In reference to the 'Green Line' on Israeli maps which identify 'Cease-Fire Lines'. The term was later adopted by UN observers in Cyprus. Beirutis explain the name by the luxuriant vegetation growing in the no-man's land between the two sectors of the city, thriving on burst water-pipes and drains.
2 The Beiruti Greek-Orthodox (known as 'Roum') are of course Arab; the term 'Greek' refers to the language of the liturgy and not to any ethnic origin. They have been part of the Patriarchate of Antioch since the early days of Christianity.
3 The slopes of Mount Hermon (Jabal al-Sheikh) to the south-east of today's Lebanese Republic.

REFERENCES

Beyhum, N. (1991) 'Espaces éclatés, espaces dominés: étude de la recomposition des espaces publics centraux de Beyrouth de 1975 à 1990', Unpublished thesis, Université Lyon 2-Lumière: Faculté de Sociologie.

Bourghey, A. (1970) 'Problèmes de géographie urbaine au Liban', *Hannon* (Beirut), vol. V.

Buheiry, M. (n.d.) 'Beirut's role in the political economy of the French Mandate, 1919–39', *Papers on Lebanon* (Oxford), 4.

Chakhtoura, M. (1978) *La guerre des graffiti: Liban 1975–78*, Beirut: Dar al-Nahar.

Davie, M.F. (1983) 'Comment fait-on la guerre à Beyrouth?', *Hérodote* (Paris), 29–30.

—— (1984) 'Trois cartes inédites de Beyrouth. Eléments pour une histoire urbaine', *Annales de Géographie de l'USJ* (Beirut), vol. 5.

—— (1987) 'Histoire démographique des Grecs-Orthodoxes de Beyrouth, 1870–1939', Unpublished mémoire de maîtrise, Beirut: Université Saint-Joseph.

—— (1992) 'Beyrouth "Est" et Beyrouth "Ouest". Aux origines de clivage confessionnel de la ville', *Les Cahiers d'URBAMA* (Tours), vol. 7.

Gedeon, C. (1974) 'Les six camps de réfugiés palestiniens dans l'agglomération de Beyrouth', Unpublished mémoire de maîtrise, Beirut: Ecole Supérieure des Lettres.

Ghorra, C. (1975) 'Les centres résidentiels suburbains dans la croissance de Beyrouth', Unpublished thesis, Toulouse: Université Toulouse-Le Mirail.

Ghorra-Gobin, C. (1983) 'Le processus de création des "Centres Résidentiels": une greffe de modernité dans l'agglomération de Beyrouth?', *Annales de Géographie de l'USJ* (Beirut), vol. 4.

Habib, T. (1986) 'Les graffiti de la guerre libanaise (1975–1983). Analyse socio-politique', *Annales de Sociologie et d'Anthropologie de l'USJ* (Beirut), vol. 2.

Johnson, M. (1986) *Class and Client in Beirut. The Sunni Muslim Community and the Lebanese State, 1840–1985*, London: Ithaca Press.

Kereughlian, A.A. (1970) 'Les Arméniens de l'agglomération de Beyrouth. Etude humaine et économique', Unpublished mémoire de maîtrise, Beirut: Ecole Supérieure des Lettres.

Khalaf, S. (1979) *Persistence and Change in 19th Century Lebanon. A Sociological Essay*, Beirut: American University of Beirut.

Khuri, F. (1975) *From Village to Suburb. Order and Change in Greater Beirut*, Chicago: University of Chicago Press.

Kongstad, P. and Khalaf, S. (1974) *Hamra of Beirut. A Case of Rapid Urbanization*, Leiden: Brill.

Longrigg, S. (1958) *Syria and Lebanon under French Mandate*, London: Royal Institute of International Affairs. (Reprint, Beirut: Librarie du Liban.)

Nasr, S. (1985) 'La transition des Chiites vers Beyrouth: mutations sociales et mobilisation communautaire à la veille de 1985', in *Mouvements communautaires et espaces urbains au Machreq*, Beirut: Cermoc.

Owen, R. (1981) *The Middle East in the World Economy, 1800–1914*, London: Methuen.

Phares, J. (1973) *Une société banlieusarde dans l'agglomération de Beyrouth*, Beirut: Université Libanaise.

Quabain, F.I. (1961) *Crisis in Lebanon*, Washington: Middle East Institute.

Schatkowski-Schilcher, L. (1985) 'Families in politics. Damascene factions and estates in the 18th and 19th centuries', *Berliner Islam Studien* (Berlin), vol. 2.

Seurat, M. (1985) 'Le quartier de Bâb Tebbané à Tripoli Liban: étude d'une 'assabiyya urbaine' in *Mouvements communautaires et espaces urbains au Machreq*, Beirut: Cermoc.

Smith, P.A. (1986) 'The Palestinian diaspora, 1948–1985', *Journal of Palestine Studies* XV(2), 59: 90–108.

Tabet, J. (1986) 'Beyrouth et la guerre urbaine: la ville et le vide', *Peuples méditerranéens* (Paris) 37: 41–9.

Tarazi-Fawaz, L. (1983) *Merchants and Migrants in Nineteenth-century Beirut*, Harvard: Harvard University Press.

4

THE PALESTINIAN STATE
Options and possibilities
Ghada Karmi

INTRODUCTION

This chapter was written during the summer of 1991, before the 'peace process' began in Madrid and well before the peace accord was drawn up between Israel and the Palestine Liberation Organisation. It reflects the major concerns of the time which were with finding a realistic, if not wholly equitable, solution to one of the longest running conflicts of our time. The second Gulf War had recently ended and had opened a window of opportunity for a resolution to the Palestine problem through President Bush's new world order. While the Arab–Israeli peace process, commencing with the Madrid conference of October 1991, did indeed result from this new approach, the outcome by August 1993 was highly disappointing. In particular, the bilateral talks between Israel and the Palestinians yielded no concrete results, and there were indications that the whole process might be suspended. It was at this point, at the end of August 1993, that news of a separate and secret peace agreement between Israel and the PLO broke, and the situation took on a radically different complexion.

In recent years, the solution to the Palestine problem has come to be seen as hinging on the creation of a sovereign Palestinian state, to be established alongside Israel, on territory vacated by Israel. This was conventionally accepted to mean the West Bank and Gaza with East Jerusalem as its capital, and it was argued that this two-state solution represented the only viable option, since it offered the Palestinians their minimum demands for self-determination and an end to military occupation, while giving Israel the lion's share of the original Palestinian territory within secure borders. Nor was this solution without its critics on the Palestinian side, since they argued that, even if it was realistic, it was still inequitable. At the same time, the official Israeli view has been

59

Table 4.1 Palestinian populations in the Middle East, 1989

Country	Population ('000s)
Israel and the Occupied Territories:	
Israel (pre-1967 borders)	800
West Bank	900
Gaza Strip	550
Total	2,250
The Arab world:	
Jordan	1,700
Kuwait (pre-August 1990)	400
Lebanon	350
Saudi Arabia	250
Syria	225
Iraq (pre-August 1990)	70
Egypt	60
Libya	25
Others*	425
Total	3,505
Overall total	5,755

Source: Extrapolations from Blake and Drysdale 1985: 284.
Note: * Includes 165,000 in Europe and the USA.

to reject a Palestine state under any circumstances and to resist an exchange of Palestinian land for peace.

When these proposals are compared to those offered by the peace deal today, they seem almost utopian. For the Gaza–Jericho First Agreement only allows for limited Palestinian self-rule in both these areas, a re-deployment but not withdrawal of the Israeli army in the Occupied Territories, and excludes Jerusalem altogether. Many Palestinians feel cheated and short changed, and that Israel has given little and gained much. Nevertheless, the agreement does provide for future developments, albeit in phases, and it is undoubtedly the potential rather than the actual terms of the agreement which have drawn so much Palestinian support. It is no secret that many Palestinians (and Israelis as well) believe that the present accord will ultimately lead to the creation of a Palestinian state, whatever the official Israeli opposition. If so, then there are practical and theoretical issues surrounding the implementation of the Palestine state which need to be addressed.

This chapter does not provide a blueprint for the Palestine state, but sets out for discussion the major issues which will inevitably arise around a two-state solution. I make no apology for the references it makes to Palestinian rights and principles which some may consider that the present peace agreement has rendered obsolete.

DEFINITIONS AND PRINCIPLES

Any discussion of solutions to the Palestinian problem, if it is to be fruitful, must begin from a common set of definitions and principles. This chapter, therefore, first addresses itself to these definitional problems:

1 The term 'Palestinians' includes those who are resident within the Israeli-occupied territories of the West Bank (including East Jerusalem) and the Gaza Strip, as well as those in exile outside these territories and Israel itself. It must also reflect the concerns and interests of those Palestinian residents of Israel itself, although they have Israeli citizenship. Any solution to the Palestinian problem must therefore take the needs and interests of all these Palestinians into account. Although it is not suggested that the issue of the Arab population should result in any attempt to undermine Israeli sovereignty, such persons have rights to national recognition within a Palestinian context as well.

2 These issues of definition are important because a tendency, chiefly encouraged by Israel, has developed which considers the Palestinian problem as being confined only to the inhabitants of the West Bank and the Gaza Strip. Yet, just as world Jewry has often viewed Israel as a home for all Jews, so Palestinians regard the whole of Palestine as a home for all Palestinians. This chapter therefore discusses possible solutions to the Palestinian problem which are directed at all Palestinian people, wherever they live – and the majority reside outside the Occupied Territories (see Table 4.1).

Quite apart from the definitional problem, there is also an issue of what the principles involved in the discussion really are:

1 The Palestinian problem is concerned with the rights of a 'people'. A solution to the Palestine problem must be based on those inalienable rights and principles which are recognised by the world community and formulated in the United Nations Charter. It also follows that, while the mode of implementation of these principles

could be a matter for negotiation, the principles themselves cannot be subject to negotiation.

2   At the same time, the principles involved must be applied in a way that is acceptable to all parties involved in the negotiations. Clearly, any peace settlement will need to be structured so as to ensure the development of mutually viable relations between Israelis and Palestinians. This will mean a relationship of economic as well as political equality, not one based on Palestinian dependence on a powerful Israeli economy. For that reason, there will have to be an interim period involving international guarantees while Palestinians establish their own institutions and structures free of Israeli control. This would then be followed by the normalisation of diplomatic and economic relations on an equal basis.

3   The principle of reciprocity and balance is essential in formulating any peace settlement. This means that aspects of the settlement can only be effectively negotiated between the parties on a reciprocal basis. Exact symmetry is not necessary but neither party's proposals should be accepted at the expense of those formulated by the other party.

4   Any peace settlement must also include the basis for future co-operation between the sovereign entities it defines. A peace settlement based on territorial partition between Israel and the Occupied Territories, wherever the final borders between them may be, should aim to facilitate future economic linkage between the two entities, rather than separate development. Economic and resource issues, as well as population settlement or residence and citizenship rights, cannot be resolved in isolation.

INALIENABLE RIGHTS

In common with all other peoples, as defined by the United Nations Charter, Palestinians have inalienable national and human rights which cannot be the subject of negotiation. These must be fully and completely respected in any negotiation of a solution to the Palestinian problem.

National rights

Palestinians have absolute and inalienable rights of self-determination and statehood. Discussion on this point must therefore be directed towards the creation of a Palestinian state.

Human rights

Palestinians have absolute and inalienable rights as defined in the United Nations Charter of Human Rights. They have the right to live free from military occupation (except as provided for under the Geneva Conventions of War), physical attack, enforced exile and harassment. Palestinians have the natural right to reside in their place of origin – even though their current place of residence or birth may differ as a result of enforced exile.

TOWARDS A SETTLEMENT

The negotiation of a settlement to the Palestinian problem in terms of these two principles will require detailed consideration of several different issues:

Palestinian representation at the negotiating process

As far as the Palestinians are concerned, the Palestine Liberation Organisation (PLO) – as the legitimate representative of the Palestinian people, as laid down in May 1964 by the Palestinian National Council (PNC) and as confirmed in the 1974 Rabat Arab League Conference – remains the only possible ultimate negotiating partner. However, if this is unacceptable to the other party in negotiations, the PLO will not object, in principle, to elections taking place to select a political leadership, provided the following points are observed:

1 Such elections would have to be conducted both in the Occupied Territories and amongst those Palestinians in enforced exile. It is appreciated that a complete electoral process involving the latter group would be extremely difficult to organise; none the less, serious attempts should be made towards their inclusion. For example, those Palestinians registered with UNRWA (United Nations Relief and Works Agency) as refugees could participate.
2 The elections would have to be conducted under United Nation auspices, as is usual in matters pertaining to national self-determination. Guarantees of immunity from harassment and arrest would have to be provided for the safety of those elected.
3 The electoral constituency must include the residents of East Jerusalem, currently illegally annexed to Israel.

The rights of elected members

Once the elections have been completed, certain additional principles must also be observed:

1   The other party to negotiations may not exclude elected members on the grounds that they also are members of the PLO or its associated organisations.
2   Those elected from the Occupied Territories must be freely able to attend meetings of the PNC – effectively the Palestine parliament in exile – and to take part in the decision-making process there, along-side other Palestinians elected from the Palestinian communities still in enforced exile. Nor shall they be barred from publicly supporting such decisions within the context of any negotiating process.
3   Notwithstanding the elections, a Palestinian government-in-exile may also be constituted. This may be achieved through the PNC, with or without an electoral process, and shall be accorded full rights of recognition. In this context, it shall be entitled to operate as the provisional government of a future Palestinian state. Any such government shall take at least half of its members from amongst those resident in Occupied Palestine. Such persons may include those illegally deported from the Occupied Territories since 1967.

ELEMENTS OF A SETTLEMENT

Any eventual settlement must resolve three basic issues – the nature of the Palestinian state; the status of Israel's occupation of areas outside its pre-1967 borders; and the future of Jewish settlements within the territory of a future Palestinian state.

The Palestinian state

The boundaries of the state

These are to be based on the terms of United Nations Security Council Resolution 242 (1967) which, 'Emphasising the inadmissibility of the acquisition of territory by war' requires the 'withdrawal of Israeli armed forces from territories of recent conflict' and affirms the necessity for 'a just settlement to the refugee problem' and 'the territorial inviolability and political independence of every state in the area'. The resolution has been accepted by the world community and has been re-affirmed by President Bush in the aftermath of the recent war in the Gulf.

Palestinian acceptance of this resolution represents a considerable concession, since United Nations Security Council Resolution 181 (1947) on the partition of Palestine has still not been implemented. Had it been implemented, it would have given the Palestinian state created thereby a far greater proportion of the available land and, since it has never been contradicted by later United Nations resolutions, it still represents the official view of the World Community of Nations on what would have been an equitable solution of the issue of the partition of Palestine. None the less, Palestinians are prepared to negotiate on the exact boundaries and recognise that it may be necessary to respond to Israeli security concerns. Three points, however, must be made in this connection:

1 The Palestinian requirement for the inclusion of East Jerusalem in the future Palestinian state is not open to negotiation. The Israeli annexation of the city remains illegal and will be treated as such in any negotiation.
2 Territorial adjustments to proposed border arrangements can be made, provided only that the principle of reciprocity is applied. This means that where specific areas of land are conceded from the Palestinian state through negotiation, equivalent areas of territory must be conceded from land currently construed to form part of the territory of Israel within its pre-1967 borders.
3 A territorial corridor under Palestinian control, linking the West Bank and the Gaza Strip, will have to form part of any territorial settlement and boundary delimitation.

Security issues

It is recognised that it will be necessary to address the issues of de-militarisation and buffer zones in any negotiation. Genuine and serious Israeli concerns about security and the potential military threat of a Palestinian state can be met by demilitarisation of the state for a specific period, to be agreed through negotiation. The same principle would apply to the alternative of United Nations' peace-keeping forces in buffer zones for those regions considered by Israel to be 'sensitive' in security terms. However, the Palestinian state, in its turn, will also require guarantees – underwritten by the United Nations or the United States of America – from Israel which would satisfy its own security concerns.

The right of return

If the negotiators agree that the Israeli 'law of return' should continue to be applied in Israel, within its pre-1967 borders, then there will also have to be a corresponding agreement by Israel on the recognition and fulfilment of United Nations Security Council Resolution 194 (1948), which recognised the right of Palestinians to return to Palestine.

In so far as the full implementation of this resolution is not practical in current circumstances, an agreement over appropriate compensation for Palestinians who have lost property and property rights in Israel will have to be made.

Alternative arrangements to compensation may be considered. These could include mutual rights of settlement for Israelis and Palestinians in each other's states with the provision of citizenship in either state for those wishing to reside there.

Natural resources

Water and fertile land in the Occupied Territories are very scarce as far as the Palestinian population there is concerned.

Although Israel's population is only 66 per cent of the total population of the pre-1967 state and the Occupied Territories, it consumes 86 per cent of all the water. In the Gaza Strip, per caput Israeli water consumption is seventeen times greater than that of the per caput Palestinian consumption (2,000 cubic metres per Israeli in 1984, compared with 115 cubic metres per Palestinian). In the West Bank the corresponding ratio is 4:1 in favour of Israeli settlements (Petch 1990).

The situation as far as land is concerned is similar. Between June 1967 and June 1968, Israel took over 700,000 *dunams* (around 175,000 acres) of Jordanian state land and 320,241 *dunams* of abandoned property (Nazzal 1980). Between 1967 and 1979, 61,000 *dunams* of West Bank land and 500 *dunams* of Gaza land were expropriated by Israel for military purposes. Purchases by Israelis of privately owned land during this period totalled 80,000 *dunams* after 1979 (when such purchases were permitted, despite the requirements of the Geneva Conventions). A further 430,000 *dunams* in the West Bank and 8,000 *dunams* in Gaza were taken over under 'Custodian of Absentee Property' legislation, while 1.53 million *dunams* in the West Bank and 63,000 *dunams* in Gaza were said to be of unclear title. Only 200,000 *dunams* in the West Bank and 253,000 *dunams* in the Gaza Strip were owned by Palestinian residents there (Abu-Lughod 1982, Gould 1991: 113).

It is clear that, in this context, equitable agreements will have to form part of any final settlement over the use of land and water resources between the two states, taking into account basic requirements for domestic use, industrial use and agriculture.

Economic and cultural links

These will require detailed attention, as was the case with the division of land and water resources. Agreements over such links should also allow for reciprocal and equivalent rights of access to each state for those wishing to work and for cultural and religious reasons.

Political representation

Jews or Israelis who are prepared to take citizenship within the Palestinian state should ultimately be entitled to appropriate proportional representation within the political institutions of the state. However, this concession must be paralleled by the provision for appropriate political participation of the Palestinian population of Israel – something which does not apply at present, where the 17-per-cent-strong non-Jewish community is vastly under-represented in the Knesset.

The military occupation

The Israeli military occupation of the West Bank and the Gaza Strip should end as the first step of any negotiating process directed towards a solution of the Palestinian problem. Israeli forces should be replaced by United Nations units for an interim period while the ultimate political settlement is established.

Jewish settlements

Jewish settlements inside the Occupied Territories were established illegally and against the provisions of Article IV of the Geneva Conventions. They should, therefore, be disbanded. The settlers themselves should leave the Occupied Territories. Jews who apply for citizenship inside the Palestinian state would acquire rights of residence, as provided for above, and, thereby, the opportunity to acquire property. However, title in this case cannot be based on the illegal title granted to current Jewish settlements.

RECENTLY PROPOSED PEACE PLANS

Recently, a number of peace plans have been put forward, all advocating in one form or another the creation of a Palestinian state to be based more or less in the Occupied Territories. I have highlighted below the main points of two of the most interesting ones.

Peace in stages

This plan was put forward by Shmuel Toledano, a former Knesset member, head of intelligence, and Israeli government adviser on Arab affairs. It is said to have attracted wide support among Israelis. It provides for a graded five-year programme, leading to the creation of a Palestinian state in the territories occupied by Israel after 1967, excluding Jerusalem and the neighbourhoods built around it by Israel after that time. During the five years before Israel vacates the territories, there are a number of conditions which the Palestinians and Arab states have to fulfil, including recognition of Israel's right to existence, a halt to the *intifada*, and the revoking of the Palestinian right of return to pre-1967 Israel. After its creation, the Palestine state is to have only a symbolic army for a period of ten years. Israel will have the right to conduct reconnaissance flights over the new state for the first five years. The main merit of this plan, which seems to be aimed almost exclusively at allaying Israeli security anxieties, is that it does provide for an unambiguous withdrawal from the Occupied Territories and an acceptance of the creation of a Palestinian state with PLO participation. Its major weaknesses, however, lie in its retention of Jerusalem under Israeli sovereignty and the maintenance of Israeli military occupation in the territories for a further five years, and terminated even then only subject to Palestinian 'good behaviour'.

Arab–Israeli peace, a new vision for a transformed Middle East

This plan was drawn up by Muhammad Rabie, a Palestinian, formerly Professor of Political Science at Kuwait University, and currently the director of the Centre for Educational Development in Washington. It provides for a more comprehensive settlement to the conflict and draws in other Arab states, particularly Jordan, which is given a major role. Pre-1948 Palestine would be designated as a shared homeland between Israelis and Palestinians, each in their own state. The borders of these states would be roughly those of the pre-1967 partition lines. The right of return and citizenship would apply for each people to its own state only, but rights of residence (which, however, exclude political rights) may be granted on a reciprocal basis. Jerusalem would be the capital of both states and governed by a joint municipal council.

Further arrangements would be concluded between the Palestinian state and Jordan, between Israel and Jordan, and between all three. In the first case, the territories would be merged to form one Jordanian–Palestinian homeland with a common (Jordanian) army for defence. In the second case, security and defence agreements would be reached between Jordan and Israel and a multi-national force would be stationed along the Israeli–Palestinian and Israeli–Jordanian borders. Thirdly, all three states would form an economic union to address such issues as trade, labour and investment.

Finally, the plan envisages an arrangement between Israel and the Arab states. Israel would withdraw from the Golan Heights, which would revert to Syria, but there would be a multi-national force stationed along the Syrian–Israeli border. Likewise, Israel would withdraw from Lebanese territory and then other non-Lebanese forces would likewise withdraw. The creation of a Middle Eastern Economic Community (MEEC) composed of Egypt, Iraq, Israel, Jordan, Lebanon, Palestine and Syria, would serve as a vehicle for regional cooperation leading ultimately to economic integration and a common market. Like the European model, it would eventually enable the free movement of goods, capital and people between member states. A regional conference along the lines of the Conference on Security and Cooperation in Europe (CSCE), to be called the Middle East Conference on Security and Cooperation and to include all Arab states, Israel, Turkey, Iran and Ethiopia, should be convened as a permanent conference. It should endorse and coordinate efforts to deal with the region's other problems, including border disputes, arms spending, water resources and human rights in member states.

This is an interesting plan because it has a long-term vision, aims to resolve all sources of actual and potential conflict on a regional, rather than on a narrow local basis, and is the first which I am aware of to transform Israel into a Middle Eastern state. Some may view it as utopian, given the lack of political will both in Israel and the United States to find a comprehensive settlement. It will not satisfy all Palestinians or Arabs either, since it involves the renunciation of many hopes and dreams. It has nothing to say about the difficulties of union between states with such differing political systems as Israel and the Arabs. Nevertheless, it deserves to be considered as a basis for discussion towards a settlement.

Whether any of these plans is likely to succeed is uncertain. Their main significance lies in the fact that they are being put forward, not by governments and official bodies, but by individuals hoping to fill the vacuum which exists in official international thinking on the question of peace between Israel and its neighbours.

REFERENCES

Abu-Lughod, J. (1982) 'Israeli settlements in occupied Arab lands: conquest to colony', *Journal of Palestine Studies*, Winter.

Blake, G.H. and Drysdale, A. (1985) *The Middle East and North Africa: A Political Geography*, New York and Oxford: Oxford University Press.

Gould, St. J.B. (1991) 'Jewish settlements and land in Israel's Occupied Territories', in N. Bexhorner and St. J.B. Gould (eds) *Altered States? Israel, Palestine and the Future*, London: SOAS, London University.

Nazzal, N. (1980) 'Land tenure, the settlements and peace', in E.A. Nakhleh (ed.), *A Palestinian Agenda for the West Bank and Gaza*, Washington: American Enterprise Institute for Public Policy Research.

Petch, J. (1990) 'Water in Israel and the Occupied Territories', in N. Beschorner and St.J.B. Gould, *Altered States? Israel, Palestine and the Future*, London: SOAS, London University.

Peace Plans

Texts of these plans may be obtained as follows: *Peace in stages*, c/o YAKAR, 2 Egerton Gardens, London NW4 4BA. *A Vision for the Transformation of the Middle East*, direct from the author, Dr Muhammad Rabie, Centre for Educational Development, PO Box 25309, Washington, DC 20007.

5

THE FUNCTIONAL PRESENCE OF AN 'ERASED' BOUNDARY

The re-emergence of the Green Line[1]

David Newman

INTRODUCTION: FORMAL AND FUNCTIONAL ASPECTS OF BOUNDARY STUDIES

Functional analyses of international boundaries have focused on the spatial impact of the boundary on its surrounding region – hence defining the existence, and extent, of a political frontier. The nature of relations between neighbouring states is generally seen as determining the location of boundaries along a closed–open continuum, thus governing the degree of contact and interaction (social, economic, cultural) between the residents of the two sides (East and Prescott 1975, Prescott 1987, Newman 1989). Political scientists and geographers have focused on the boundary 'opening' process which has taken place in recent years, resulting both from a decrease in political tensions in some parts of the world (such as Western Europe) as well as the fact that the global village communications network has destroyed one of the major 'sealing' characteristics of international boundaries.

As such, boundary studies assume a somewhat deterministic position in that the political status of a boundary as determined by government ('closed–open') largely determines the nature of relations between the populations on both sides. This is especially true of 'closed' boundaries, wherein governments supply the physical means by which actively to prevent cross-frontier contact. But in reality, the formal and functional characteristics of boundaries feed on each other within a system of mutually enforcing influences (Figure 5.1). Depending on the unique circumstances of each boundary, parallel processes of contact and separation may take place at one and the same time. This may often reflect

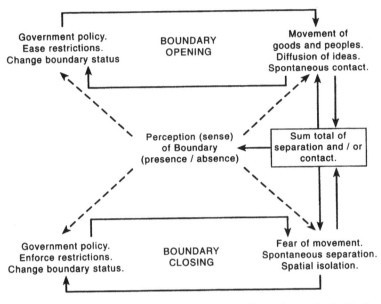

Figure 5.1 Formal and functional characteristics of boundaries: the feedback effect

differences between governmental (formal) and frontier residents' (functional activity patterns) perceptions of the significance of the boundary within the national or regional landscapes respectively. In discussing recent changes in Eastern Europe, Foucher (1990) argues that the change in the *status quo* (formal status) of the inter-state boundaries is largely due to the alterations of the functionality of these same borders. Thus in many cases it is the changing functional nature of cross-frontier relations which has often been a catalyst in forcing government to further relinquish stringent boundary controls. At the same time, while telecommunications and satellites have indeed brought about a transformation in the extent to which boundaries can retain completely 'sealed' characteristics, it still requires the physical opening of borders by governments for full contact potential to be realised. Alternately, governments are able to enforce tighter restrictions along boundaries, thus strengthening processes of separation. Indeed, while scholars have been focusing on recent processes of boundary 'opening' and 'removal', political events elsewhere in the world are bringing about the creation of international boundaries between newly emerging states. The functional characteristics of these boundaries have yet to undergo formative processes, and much will depend on the nature of ethno-political relations between the states as well as the degree to which existing economic links and mutual dependencies need to be retained. Thus, while many 'closed' boundaries are gradually opening up in the face of political integration processes, other previously 'open' boundaries (which were previously no more than administrative boundaries) may take on more 'closed' characteristics, serving as new barriers to communication and trade.

Administrative boundaries have an important role to play in this process of international boundary evolution. It has generally been assumed that an administrative boundary is no more than an intra-state division, of little importance in the study of inter-state relationships. But administrative boundaries which separate distinct ethnic regions can speedily be transformed into international boundaries under a changed system of power relations. Recent events in the Soviet Union and Yugoslavia offer clear evidence of this process. The imposition of some form of political union by a central authority – rather than union by consent – has not been shown to promote meaningful socio-cultural integration between ethno-territorial groups. The administrative boundary, while not displaying the blatant 'barrier' characteristics of an international boundary, continues to provide important ethno-territorial demarcators. The boundary exists for numerous daily functional activities – it is just a matter of changing its formal status should the political situation so dictate. The extent to which administrative boundaries are perceived as constituting points of functional separation and differentiation are an important part of this process. Foucher (1990) notes the importance of examining the 'boundary perceptions related to territory'. This may include 'hypothetical boundaries used as a basis for territorial claims' as well as 'erased frontiers'. The latter includes boundaries which may have formally been removed by governments (following territorial conquest and/or expansion) but which have been replaced by administrative boundaries and/or remain deeply inscribed in the perceptions of the frontier population. The perception, or 'sense' of boundary – despite its physical absence – is similar in many respects to the territorial processes observed by urban geographers when studying metropolitan and neighbourhood behaviour. Urban geographers have noted the tendency for 'invisible' boundaries to be clearly demarcated within the perception of the local residents. Lynch (1960) described these phenomenon as 'edges', constituting more or less penetrable barriers, or lines along which neighbourhoods are related and joined together. Bounded territories and feelings of territorial belonging at the local level are a microcosm of similar processes relating to international boundaries and territories. Thus we return to Soja's (1970) *The Political Organization of Space* in which the function of the boundary as a point of separation/contact becomes a more important study unit for territorial behaviour than the scale – spatial (national/local) or political (inter-state/intra-state) – at which it operates.

Contextually, the case of the 'Green Line' boundary – which separated Israel from the West Bank during the period 1948–67 – provides an interesting case-study of the interplay between intra-state administrative and inter-state national relationships. In terms of its formal (international) classification, this boundary has been in a 'non-existent' or 'erased' state since its removal at the onset of the Six Day War of June 1967. Yet a functional and/or perceptual analysis of this 'non-existent' border today would suggest that the boundary is clearly present. This is reflected both in terms of daily activity patterns on the part of both Palestinians and Israelis, as well as maintaining a clear perceptual presence in the minds of both peoples. This 'presence' of boundary has increasingly come to reflect the functional realities governing the political relations between Israelis and Palestinians, and is clearly part of a gradually evolving process of Palestinian state formation. This chapter focuses on an analysis of the way in which this 'non-existent' boundary has functionally re-emerged, despite its formal absence.

STAGES IN THE EVOLUTION OF THE GREEN LINE BOUNDARY

The Minister of Housing, Ariel Sharon, members of the Knesset ... were present yesterday at the cornerstone ceremony for a new neighbourhood in the Katzir settlement ... Katzir is one of the 'cluster' settlements which is being located in the Wadi Ara' region and along the route of what the Housing Minister described as 'having once been the green line, but which no longer exists'.

(*Ma'ariv*, 28 June 1991: 9)

Israel's control of the West Bank and Gaza Strip has extended since 1967. The West Bank, as a separate politico-territorial unit, had been in existence since 1948, following the partition of Western Palestine in the wake of the establishment of the State of Israel. Until 1967, the West Bank was governed as a territorial adjunct to the State of Jordan. Since Israel's conquest of the West Bank and Gaza Strip, the region has been subject to an Israeli military administration.

Table 5.1 Phases of 'Green Line' boundary

Period	Status	Functional characteristics	
		Separation	Contact
Pre-1948	Absence of boundary		Regional integration
1948–67	Armistice line	Spatial reorientation	
	Sealed boundary	Frontierisation	
1967–87	Boundary removal	Municipal boundaries	Palestinian labour
	Administrative boundary	Non-annexation	Settler migration
Post-1987	Administrative boundary	Geography of fear	
	Curfews and road blocks		

An examination of the history of the Green Line boundary enables us to focus on four phases of boundary evolution (Table 5.1). This ranges chronologically from a pre-boundary stage (prior to the establishment of the State of Israel in 1948) to the *de facto* re-emergence of a functional boundary (despite its formal absence) during the most recent period. The first two phases will be described only briefly, as they do not constitute the focus of this study. The third phase, describing some of the implications of territorial reintegration following the removal of the boundary in 1967, will be dealt with in greater detail. This, in turn, will be set against the fourth, and final, phase – the re-emergence of the boundary – which constitutes the major focus of the study.

Phase 1: Absence of boundary (pre-1948)

Prior to 1948, Palestine to the west of the River Jordan had never been partitioned. Indeed, the creation of Transjordan by the British Mandate was viewed by many as constituting the initial partition of Palestine into separate politico-territorial units with the boundary running in a north–south line along the course of the River Jordan. The significance of this phase of pre-partition was that the area between the River Jordan in the east and the Mediterranean Sea to the west functioned as a single spatial unit for its Arab and Jewish residents respectively. Transportation arteries, trading patterns and cultural links resulted from a spontaneous pattern of regional development throughout the entire region. As such, the international focus for the local economy – inasmuch as international trading links existed – focused on the west and the Mediterranean ports, especially Jaffa. Arab/Palestinian inter-settlement activity patterns included the whole of Palestine west of the River Jordan. Despite this, the future emergence of the 'Green Line' was already being demarcated during this period. The nature of Jewish/ Zionist settlement activity during the pre-State period dictated the settling of lowland and relatively unpopulated areas. The mountain interior of Palestine did not fit any of these criteria. As a result, Zionist settlements were not founded in what was later to become the West Bank (with the exception of the Etzion sub-region to the south of Jerusalem). Israel's 1948–9 War of Independence concentrated on retaining sovereign control over the existing Jewish settlements and in ensuring sufficient land for future development policies. Implementing the latter objective resulted in the Israeli Army moving south and capturing the Negev region, rather than moving westwards to the River Jordan. Thus, even in the pre-boundary phase, the demarcation of the future boundary was already beginning to emerge. A map showing the location of the Jewish settlement of that period clearly depicts the outline of the future boundary, beyond which hardly any Jewish settlement had been established.

Phase 2: Presence of boundary (1948–67)

Following the cessation of armed conflict between Israel and her Arab neighbours in 1949, armistice lines were demarcated. The presence of the 'Green Line' boundary took on all of the classic characteristics of a 'sealed' frontier, with differential developmental policies and spatial re-orientation taking place on both sides of the boundary. According to Brawer (1984: 160), 'the imposition of a totally sealed new boundary led to rapid frontierisation processes over a narrow strip of territory on both sides'. In his post-1967 studies, Brawer (1984, 1990) has clearly depicted the impact of this nineteen-year boundary on patterns of development, especially with regard to processes of frontier zone development policy on the Israeli side as compared with depopulation of the frontier region on the Jordanian side; differences in settlement patterns and socio-economic structures, standards of living, agricultural occupations and other demographic characteristics amongst the region's Arab population. While much of this differentiation resulted from the respective policies of governments towards this boundary region, the arbitrary division of the region – often separating villages from their lands and water resources – also played a major part in the emergence of the unique spatial characteristics of the new frontier zone (Brawer 1990).

Allowing for the differences in scale and cultures, the effect of the imposition of the 'Green Line' boundary was similar in many respects to that of the 'iron curtain' in Europe. While Palestinians within the West Bank found themselves residing on the periphery of the State of Jordan with no access to the west, those remaining within Israel found themselves slowly swallowed into the newly emerging market economy. No informal contact took place between the Arab populations on either side of the boundary, while even formal contact was limited to the absolute minimum. The reorientation of the transportation and economic systems led away from the boundary – as a national periphery – and in opposite directions into the respective state centres. This was much more marked on the West Bank side of the boundary owing to the real distance between the political frontier and the national centre. On the Israeli side of the boundary, large parts of the 'Green Line' were located close to the Israeli metropolitan centres, allowing for development to take place in parts of the frontier region. Within the wider region, the sudden imposition of a 'sealed' boundary influenced the geographic development of several towns, such as Haifa and Ashdod (port towns) in Israel, and Gaza and Hebron located beyond the 'Green Line' (Drysdale and Blake 1985: 104).

Phase 3: Removal of boundary (1967–mid-1980s)

The outcome of the 1967 War was effectively to remove the presence of the 'Green Line' boundary. According to Brawer (1984: 160): 'After functioning as a barrier of extreme separation for nearly two decades, it abruptly lost its previous character and became an internal administrative boundary, setting into motion rapid "de-frontierisation" processes in the frontier zone.'

This process of 'de-frontierisation' resulted from both formal and spontaneous functional processes, respectively feeding on each other. At the formal level, Israeli governments have continuously stated that the 'Green Line' is an artificial geographical phenomenon belonging to the past. One of the most blatant policies designed to emphasise this point was the removal of the 'Green Line' from all official maps used within Israel, including school textbooks. Israeli children born into the post-1967 geo-political reality have largely been ignorant of the presence of the 'Green Line' boundary and, until recently, would have been hard put to locate such a boundary on a map of the country (Newman 1990). At the same time, Palestinian maps also tend to depict the region as a single territorial unit, largely ignoring the prior existence (albeit for only a short period of time) of a political division. This mutual ignoring of any form of political boundary brings the Israel–Palestine conflict into sharp focus – each side is attempting to gain ultimate control of the whole territory lying between the River Jordan and the Mediterranean.

The opening of the boundary, despite the asymmetrical set of political relations, has brought about a certain amount of economic integration between the two territories with a significant movement of both goods and people in both directions. Part of this movement has been spontaneous in the sense that it arose out of mutual need of one ethnic group for the services/goods/employment offered by the other, although it could equally be argued that government policy was aimed at creating this system of mutual demand and supply as a means of 'formally' promoting territorial interdependence.

The most significant of these processes was the gradual integration of the Palestinian labour force into the Israeli market economy. Palestinian labourers were transformed into the cheap labour of Israel's economy (Semyonov and Lewin-Epstein 1987, Portugali 1991). The relative proximity of both the West Bank and the Gaza Strip to the major Israeli metropolitan centres resulted in the daily movement of tens of thousands of Palestinians into the Israeli work-force. Many of these labourers constitute the informal sector of the work-force, receiving minimal wages for their efforts, lacking the cushion of Israeli social security and unemployment benefits. For their part, many Israeli employers rely on the Palestinians for their source of cheap labour, thus reducing production costs. Despite the partial withdrawal of Palestinian labour as a result of the *intifada*, this trans-territorial labour migration continues to be an important factor governing Israel–West Bank relations.

Paradoxically, the initial Israeli policy of opening the boundary – and thus providing a new, alternative, source of employment for the Palestinians – has been a major factor causing many Palestinians to remain within the West Bank, whereas prior to 1967, the lack of job opportunities caused many more to emigrate. This is problematic for irredentist politicians who argue for the ultimate retention of the West Bank by Israel and who, by virtue of their argument, desire as small a Palestinian population as possible.

A further source of movement to have resulted from the removal of

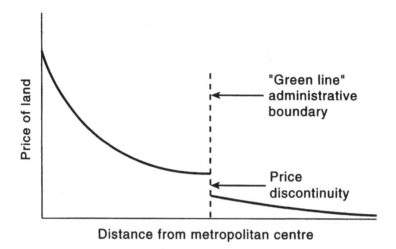

Figure 5.2 The 'Green Line' as a boundary of 'price discontinuity'
Source: After Reichman 1986.

the boundary has been the migration of Jewish settlers from Israel into the West Bank and Gaza. The process of Jewish colonisation in the West Bank has been studied in much depth (Harris 1980; Benvenisti 1984; Newman 1985, 1991 and 1992). By the end of 1991, some 120,000 Jewish settlers had moved into the West Bank (excluding a similar number in the suburbs of East Jerusalem). Well over half of these settlers had settled in the areas closest to the 'Green Line' boundary, owing to the proximity to their workplaces in metropolitan Israel. The providing of cheap land as an economic incentive to these settlers meant that the 'non-existent' boundary became an important line of 'price discontinuity' (Reichman 1986) disrupting the normal distance decay effect of the land market process wherein the greater the distance from the metropolitan centre, the cheaper the land. Once across the old boundary, land prices took a sharp dive, thus inducing potential settlers to relocate just within the western margins of the West Bank – giving rise to a process of suburban colonisation (Newman 1991, 1992) (Figure 5.2).

Another important contact characteristic which has taken place has been the 're-uniting' of Israeli Arabs with West Bank Palestinians. This is not only reflected in renewed contact between families which had been separated, but also in 'cross-frontier' marriages and educational institutions. Arab citizens of Israel have increasingly come to identify with the Palestinian cause, as the latter message has become spatially diffused across the boundary. The younger generation of Israeli Arabs perceive themselves as Palestinians, in contrast to the previous self-identification as 'Israeli Arabs'. This is clearly reflected in the trend towards the establishment of separate Arab political parties, representing Arab social, political and national concerns, as contrasted with previous periods in which a large percentage of Israeli Arabs voted for Israeli (Jewish – and even Zionist) parties.

Phase 4: *De facto* re-emergence of functional boundary

Despite the processes of territorial integration which have resulted from the removal of the boundary, a number of counter-processes emphasising the existence of the boundary have also been in operation. Of these, some have been in operation since 1967, having resulted from government policy of non-annexation of the West Bank. These processes have been reinforced by more recent phenomena which have resulted from the changed political situation in the wake of the *intifada*. As a result, the 'Green Line' remains a strongly inscribed image amongst both Israelis and Palestinians. For the former, it depicts a line beyond which a different set of laws apply and around which the Arab–Israel conflict is brought into sharp geographical focus. For Palestinians, the 'Green Line' represents the boundary of any future autonomous or sovereign territory, to be separated from Israel as was the case prior to 1967. We are able to identify five, often overlapping, factors in this process of boundary re-emergence:

The overlapping of municipal boundaries

With the exception of East Jerusalem, no part of the West Bank or Gaza Strip has been formally annexed to Israel – not even by the extreme right-wing governments of the 1980s. This non-annexation has important administrative significance for both Israelis and Palestinians. In the first place, Jewish and Arab residents of the West Bank are administered according to separate administrative frameworks, operating within the same geographical space. The Palestinian settlements are divided into six regions, mostly similar to the spatial division which existed prior to 1967 (Benvenisti 1984). For their part, the Jewish settlements are administered by virtue of their being located within one of the six regional councils which have been founded as a means of catering to the municipal needs of the new communities. While the latter are subject to Israeli planning law and legislation (albeit through the directives of the Military Administration), the Palestinian settlement framework is governed by a mixture of military and Jordanian ordinances. Within Israel proper (pre-1967 boundaries), all settlements (Jewish and Arab) form part of a single local government framework. The 'non-existent' Green Line therefore operates as a dividing line between a single local government system to the west, and a dual administrative framework to the east.

While the Jewish residents of the West Bank are administered according to their own separate system of local government, the spatial division of these administrative units nevertheless corresponds with the 'non-existent' boundary. The six regional (rural) councils within the

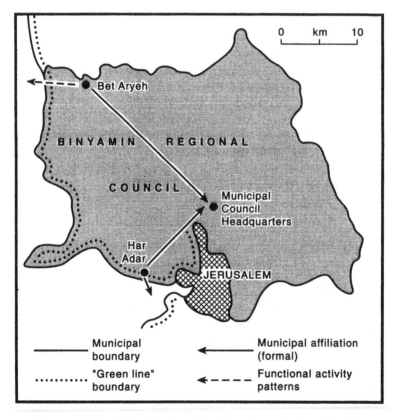

Figure 5.3 Municipal affiliation of Jewish settlements either side of the 'Green Line'

81

West Bank closely follow, as their outer line of demarcation, the Green Line. This has already caused a certain amount of administrative inefficiency, especially as regards those settlements which are located adjacent to the old boundary. Many of these suburban communities would prefer regional municipal frameworks which would enable them to function together with similar communities located within the pre-1967 boundaries – in some cases only a few miles distant. Instead, they find themselves operating within a regional council, whose headquarters are some 10–15 miles to the east, within the interior of the West Bank. Thus, instead of a natural orientation towards the west and the Israeli metropolitan hinterland, these suburban communities are forced to function as part of municipal entities towards the east (Figure 5.3).

The reason for this stringent adherence to the 'non-existent' boundary follows from the non-annexation of the West Bank by Israel. As such, Israel is unable to justify internally the extension of full civilian administrative frameworks to this region. The latter can only operate within sovereign Israel, that is, up to the old 'Green Line' boundary. This line therefore marks the point at which civilian administration ends and military administration commences. While the actual operation of these councils is the same on both sides of the 'Green Line', the West Bank councils formally operate under the jurisdiction of the military administration. In reality, the differentiation is fictitious, on paper it has important judicial significance. In effect, it further emphasises the continued existence of the 'Green Line' as a functional boundary of separation – be it only of an administrative nature.

Settlement policy along the political divide

Settlement policy has on the one hand served to emphasise the distinction between 'within the Green Line' to 'beyond the Green Line', while at the same time attempts have been made to erase the past boundary through the establishment of settlements in close proximity to the 'Green Line'. The major focus of suburban dormitory communities in the West Bank are those locations lying closest to the previous boundary. The fact that Jewish settlers cross this line in their daily commuting to and from their workplaces is largely unnoticed, especially following the construction and improvement of east–west transportation arteries linking metropolitan Israel to the West Bank. Yet these same settlers, by virtue of their residing in the West Bank, are classified as living within 'development areas' and are thus able to benefit from tax concessions and other advantages which their fellow Israelis,

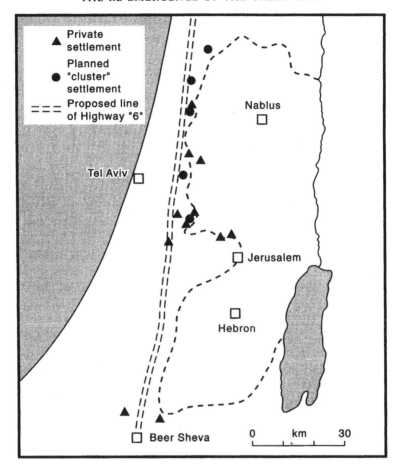

Figure 5.4 New settlement planning along the 'Green Line'

residing a few kilometres to the west (inside the 'Green Line') do not receive.

A study by Applebaum and Newman of the 'private sector settlements' which have been founded during the past decade (Applebaum and Newman 1991), also displayed an interesting locational characteristic with respect to the 'Green Line'. Unlike the majority of the dormitory communities (see previous paragraph) which rely on public sector (government) subsidies for initial infrastructural costs, the sixteen 'private sector' settlements have been founded by groups of settlers who have raised much of the necessary capital investment from their own pocket. Nearly all of these settlements are located in close proximity to the 'Green Line' but on the Israeli side of the administrative boundary (Figure 5.4). The need to attract a larger settler mass in the first instance (in order to meet scale economies), allied with the fact that the investment risk rests with the individual rather than with government, clearly dictates the choice of location within the 'non-existent' boundary. The few 'private sector' communities located in the West Bank are to be found closely hugging the 'Green Line' in areas which, it could be argued (see final section of this chapter), may be included within Israel under any future re-partition of the region.

More recently (1990) a blueprint for the establishment of seven 'cluster' townships was formally announced by the Ministry of Housing as a means of providing housing for the Israeli population during the coming decade. These 'clusters' are to be located along the route of the 'Green Line', to be linked by the construction of a new north–south national highway (Figure 5.4). The 'cluster' plan is based on the founding of twelve settlements, consisting of 28,000 inhabitants, running from Tulkarem in the north to the Ayalon Valley in the south. The government publicly perceives this, and other similar developments, as being indicative of the final demise of the 'Green Line'. Yet it could equally be argued that such a policy only goes to emphasise the line's existence, by virtue of the fact that each of the proposed settlements closely hugs the previous boundary without extending into the West Bank itself. This argument and counter-argument can be highlighted by the following extracts from the Israeli press:

> National Highway No. 6, planned to run throughout the country from south to north, closely parallels the 'green line' along much of its route. State planners are zoning land along the route for many new settlements, the majority of which are inside, but virtually touching, the 'green line'.
>
> (*Davar*, 31 August 1990: 6)

> The planners were instructed by the government not to go beyond the 'green line', but at the same time the locating of this major north–south highway was meant to be a sign that the Israeli government would never return to the 'green line' as they would not allow the international boundary to be so close to the country's major traffic artery. Others insist that the road is planned to traverse Israel through the center of the country and not along its western fringe.
>
> (*Ha'aretz*, 10 July 1990: 3)

The vast majority of the accompanying planning associated with Highway 6 are within the 'green line'. When an army spokesman was questioned concerning the minimum distance from the 'green line' within which it was permissible to build a house, the reply was '30 metres' ... In other words, we are talking about construction on the line itself and in vast quantities. In political terms, this means a strip of hundreds of thousands of new residents, all Jewish, along the 'green line'. In direct contrast to what had been previously claimed, this strip will not serve as a wedge separating Israeli Arabs from the West Bank Palestinians. Most of the Israeli Arab villages virtually touch the 'green line'. Notwithstanding, there are two approaches to the problem: One of these, which appears to have been adopted by Minister Sharon, is that this construction activity will serve to finally erase the 'green line'. The alternative approach is that the implementation of these plans will bring the presence of the 'green line' to bear in an even stronger fashion than at present.

(*Davar*, 31 August 1990: 6)

Frontier characteristics beyond the boundary

Shafir (1984) has argued that Jewish settler behaviour in the West Bank may be analysed using frontier theory, as developed by Turner (1962) on the basis of the American experience of the nineteenth century. While Turner's classic theory relates to settlement, rather than political, frontiers, the two are clearly related in the case of the West Bank. It is precisely the settlement activity aimed at meeting political objectives which transforms the West Bank into a political frontier.

Contextually, there is much to be said for the operation of frontier theory with respect to settler behaviour in the West Bank, despite the geographical proximity – and territorial contiguity – of the area in question with the political centre of the country. Many Jewish settlers have been responsible for acts of violence aimed at the local Palestinian population. This came to a height with the discovery of the Jewish underground in 1984 and again, later, with the formation of loosely organised local militias in response to *intifada* violence. But Israeli courts would appear to operate different sets of rules for Jewish and Palestinian acts of violence within the West Bank. Sentences handed down to settlers are of a relatively light nature when compared to those given to Palestinians found guilty of violence. While leaders of the Jewish underground were committed to twenty years' imprisonment,

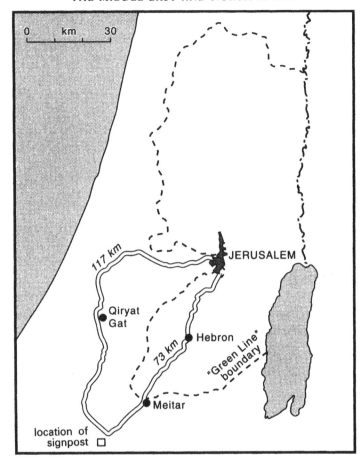

Figure 5.5 The 'Green Line' as a barrier: travelling alternatives to Jerusalem

none remained incarcerated six years later. According to a report issued by the Batzelem (human rights) organisation (reported in the *Ha'aretz* supplement, 27 December 1991: 7), out of 42 incidents in which police suspected Jewish settlers of having been involved in the deaths of local Palestinians, in only three cases was legal action actually brought before the courts. In two of these cases, settlement leaders were found guilty of serious firearm offences (one of them having caused the death of a local Palestinian), but were sentenced to relatively light punishments. Similar acts by Israelis within the 'Green Line' would have met with much heavier punishments, while out of eight similar charges brought against

Palestinians suspected of having caused Jewish deaths, seven received life imprisonment sentences (the eighth received twenty years) and their houses were destroyed. In this sense, the 'Green Line' acts as a clear boundary, beyond which there is a 'frontier' within which the often unacceptable behaviour of the Jewish settlers is indirectly backed up by an external dominant political and legal system.

Again, paradoxically, the lack of formal annexation of the region by Israel has been a major factor enabling this process to develop. Had Israel annexed the region – and extended full civilian control to the region – then both the government and the settlers would have had to operate within much more stringent legal and formal restrictions. It is precisely because the dividing line is so much more than simply an administrative boundary that different modes of behaviour – both private and public – can be practised on either side of the line.

The geography of fear – perceiving the boundary

Probably nothing has done more for the cause of the 'Green Line' than the *intifada*. Most Israelis, non-resident in the West Bank, no longer travel freely beyond the former boundary. During the period 1988–91, it became generally accepted that travelling in the West Bank was dangerous for Israelis, owing to the threat of being stoned or even fire-bombed. As a result, many Israelis became aware, for the first time, of the existence of this invisible line of separation. Figure 5.5 shows the travelling options of residents of the Metar settlement, located immediately to the south of the 'Green Line' on the road linking Jerusalem to Beer Sheba via Hebron. Until the onset of the *intifada*, it was natural for Metar and Beer Sheba residents to travel the relatively short route of 70 km to Jerusalem via the scenic routeway through the West Bank and Hebron. Since this option now presents considerable danger for the individual traveller, residents prefer to travel via the longer route, circling the West Bank altogether – a distance of some 120 km. This is of particular significance for the residents of Metar who, located 20 km to the north of Beer Sheba, even have to travel the first three km of their journey in the opposite direction to Jerusalem in order to meet up with the main highway. A road sign at the nearby Shoqet junction clearly presents the traveller with these options (Figure 5.6).

Indeed, the 'non-existent' 'Green Line' is now perceived by many Israelis as the line beyond which it is dangerous to travel. As a result, many Israelis who were previously ignorant of the precise location of the old boundary now take time to check whether planned trips involve

Figure 5.6 Road sign at Shoqet Junction showing optional routes to Jerusalem

going beyond the 'Green Line'. Many parents will not allow their children to take part in school outings if it involves passing the 'non-existent' boundary – even though, as in some cases, it may be in a peripheral region within which there is no danger of violence.

Nowhere is the 'geography of fear' felt stronger than within the 'united' city of Jerusalem. The city declared as the 'eternal undivided capital of Israel' remains physically united, but has become increasingly divided functionally. Even before the onset of the *intifada*, Israelis and Palestinians who have grown up in the post-1967 'united' phase of the city, displayed a spatial knowledge of the city which was limited to their own ethnic sector and which rarely extended 'beyond' the 'non-existent' dividing line (Romann 1989). Cross-city functional activity which did take place has largely ceased since the onset of the *intifada*, with fewer and fewer Israelis crossing the invisible divide for fear of their safety. For good reason have Romann and Weingrod (1991) described Arab–Jewish relations in Jerusalem as a case of 'living together separately'.

The perception of the 'Green Line' as a boundary of violence is well expressed in the press caricatures depicted in Figure 5.7. The first of

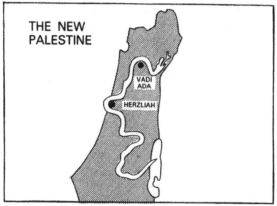

Sergeant X has been sentenced to jail for refusing to serve in the territories.

Figure 5.7 The 'Green Line' as depicted in cartoons
Source: From Shiloh: Ma'ariv.

these shows the boundary as a line of molotov cocktails, while the second and third caricatures depict the extension (diffusion) of inter-ethnic violence from the Green Line into the Arab areas of pre-1967 Israel. The second caricature appeared following an incident of stone throwing at Israeli cars within the densely populated Arab Galilee region of Israel.

Enforcing the boundary – restricting movement

The *intifada* has also brought about an increased awareness on the part of the Palestinian population of the existence of a boundary. Following acts of violence, the Israeli authorities have often imposed curfews on either the West Bank or Gaza, preventing Palestinians from 'crossing into Israel'. Since 1989, Palestinian workers in Israel have been required to obtain an identity card, without which they are not allowed to work within the Israeli cities. These cards are frequently checked by army patrols at roadblocks set up on the roads linking the West Bank to Israel, more often than not at the point of the old boundary, or in close proximity to it. The 'Erez' crossing point at the northern edge of the Gaza Strip takes on all the characteristics of a 'sealed' international boundary. Two narrow roads enable access to cars between concrete army and police checking points. This crossing point can be shut within minutes, completely sealing access to and from Israel. During Israeli holy days (such as Yom Kippur), or at times of visits by important foreign dignitaries, Palestinian access from the West Bank and Gaza is prevented altogether. During the Gulf War in January 1991, the West Bank and Gaza were sealed off for days on end. In all these cases, the 'Green Line' clearly functions as much more than a purely adminis-trative or 'non-existent' boundary.

It is, of course, paradoxical that no group has done more for the re-emerging awareness of the Green Line boundary than those who have the greatest interest in proving its demise. Following incidents of *intifada* violence, it has been the right-wing irredentist politicians who have led demands for the imposition of stricter restrictions on the free movement of West Bank and Gaza Palestinians inside Israel, while there have even been calls for the total banning of Palestinians from crossing the 'non-existent' boundary. It is precisely these same politicans who go out of their way to stress that no 'Green Line' exists in so far as an ethnic-territorial divide is concerned.

The respected military journalist Ze'ev Schiff, of the *Ha'aretz* (liberal, left-of-centre) newspaper points to this dilemma when he writes of the inherent contradiction in closing the border to Palestinians. He writes:

> the closing of the territories is an achievement for the terrorist organizations and all those who desire an intensification of the Intifadeh. It proves that ... the 'green line is alive and well' ... we can well envisage the scenario. In the next stage, there will be discussions concerning the means by which to isolate the Palestinians of the territories from the Jewish residents of the territories, and in order to avoid unnecessary loss of life, the unrestricted entry of Jewish residents from Israel beyond the 'green line' will be prevented.
>
> (*Ha'aretz*, 26 October 1990: 12)

The presence or absence of a 'Green Line' boundary, possessing more than administrative significance only, has also become part of the internal political debate within Israel. This is best depicted by recourse to the cartographic representations of this boundary. In addition to the absence of the 'Green Line' from atlases used in school (see pp. 78–9), the following two incidents serve to highlight the way by which maps are put to use in promoting the various political arguments concerning the 'Green Line':

> The Committee for Soldiers Welfare has prepared a present for all soldiers, which includes a 'touring map of Israel'. The map, prepared by the Surveyors Department of the State of Israel, has no reference to the 'green line'. This has caused much confusion amongst soldiers who have been taught to differentiate between the laws and rules applying to areas 'within the green line' and those constituting part of the 'military administration'. In a recent military exercise, soldiers were unsure as to what action to take when they were unable to ascertain which side of the 'green line' their present location was.
>
> (*Al Hamishmar*, 25 June 1986: 5)

> The Education Department of the Kibbutz Artzi Movement has been asked to prepare a map depicting the green line in time for the opening of the next school year. This is the first such map to have been prepared since 1967, since when all official maps have deleted the green line. Following the refusal of the Ministry of Housing to provide the accurate location of the green line, the Kibbutz Movement approached Professor M. Brawer of Tel Aviv University. The latter obliged by accurately demarcating the line.

The map does not include the 'green line' between Israel and the Golan Heights because, according to the Kibbutz Artzi spokesman, the annexation of the Golan Heights by Israel resulted in the annulment of any boundary in this latter region. Accordingly, pupils of the Kibbutz Artzi movement will study from maps depicting the green line as from next year.

(*Davar*, 12 June 1987)

In concluding this section of the chapter, it is evidently clear that in both the pre-1967 and post-1967 periods there has been a gap between the formal status of the boundary (presence/absence) and the functional characteristics of contact/separation. This is relevant both to Israeli–Palestinian mutual perceptions, as well as Israeli–Israeli divergent attitudes towards the political conflict in general. While a 'sealed' boundary existed between 1948–67, there was nevertheless a great deal of cross-boundary communication and awareness by one side or the other. There was no blocking of radio or television broadcasts, such that Israelis (Arabs and Jews) were always able to tune into Jordanian and other Arab state channels. Until the recent onset of an English news broadcast on Israeli public television, many English-speaking Israelis tuned into the English news broadcast on Jordanian television.

Similarly, recent years have shown that despite the absence of a formal boundary, functional characteristics of territorial separation have become increasingly stronger. This is felt equally by Palestinians and many Israelis. In essence, what we have are processes of separation and contact operating at one and the same time. Whereas the processes of contact were stronger during the period 1967–87, this has been replaced by processes of intermittent separation since the mid-1980s. These latter processes have been strong enough for a clear sense of boundary to emerge, resulting in a renewed awareness amongst both Israelis and Palestinians of the existence of separate politico-ethnic territories.

POLITICAL SCENARIOS FOR THE GREEN LINE BOUNDARY

The recent signing of an interim peace accord between Israel and the PLO raises many questions concerning the future re-partition of this region. While the 'Green Line' boundary does not necessarily have to be the precise international boundary agreed upon under any future re-partition, it clearly provides the point of departure for detailed negoti-

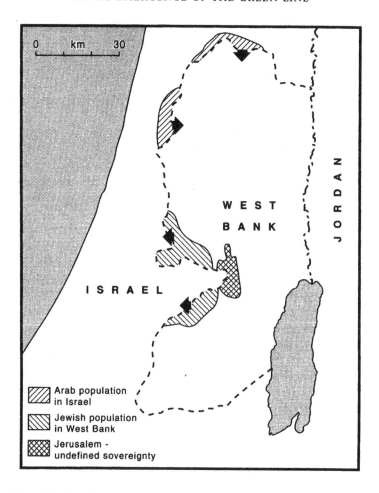

Figure 5.8 Possible territorial exchanges: re-demarcation of the 'Green Line'

ations. It may well be that the reimposition of a 'good fence' will be the first step towards the formation of 'good neighbours'.

It is not difficult to make a geographical argument for a new demarcation of a political boundary between an Israeli and Palestinian politico-spatial entity. Variations on the theme of partial territorial withdrawal have been discussed by Efrat (1982), Cohen (1986), Foucher (1987), Alpher and Feldman (1989), Kipnis (1991) and Newman (1991). These various studies present a number of territorial options, most of which would require demarcation of a new boundary between an Israeli and Palestinian political entity. Proposals for territorial re-allocation could take into account the major concentrations of Jewish settlement in the West Bank *vis-à-vis* major concentrations of Arab settlement within Israel. Three areas of Jewish settlement, each of which is located in close proximity to the prior 'Green Line' boundary could, theoretically, remain within a Jewish state, while two key areas of Israeli Arab settlement could alternately lie within a Palestinian autonomous/sovereign area (Figure 5.8). The Jewish areas consist of the Western Samaria region (the outer suburban belt of the Tel Aviv metropolitan region); the Etzion bloc of settlements to the south of Bethlehem, and possibly some of the immediate suburban colonies of Jerusalem. For their part, the large Arab concentrations of the 'Triangle' region (the towns of Tayibe and Tir'a) along the western margins of the West Bank, and some of the Wadi Ar'a settlements (the largest of which is the town of Um el-Fahm) immediately to the north of the 'Green Line' could be included within the Palestinian area. Other possible border adjustments could possibly be made in the Northern Samaria and South Hebron micro-regions owing to their geographical peripherality.

While these proposals, at this stage, sound highly speculative, the need to create regions of ethno-territorial homogeneity should not be undervalued. The bitter history of Arab–Jewish relations would indicate that, in the absence of a single bi-national entity to the west of the River Jordan, the greater the degree of territorial homogeneity, then the better the chances for some form of inter- and intra-state stability in the long term.

It is precisely for this reason that much of the most recent Israeli settlement activity – both within the pre-1967 boundaries and within the West Bank itself – is concentrated in close proximity to the 'Green Line'. The 'cluster' settlements (see pp. 85–6) located along the Israeli side of the 'Green Line' will, if constructed according to plan, provide new wedges between Arab settlements within Israel making any future boundary demarcation exceedingly difficult. Not only was the 'cluster' settlement plan intended (on the part of the planners at least) to erase the 'Green Line' (see above), through the eventual joining of the new settlements with other Jewish settlements lying just beyond the 'Green Line', but it was also intended to provide wedges preventing the territorial continuity of the Arab settlements in the Triangle region (within pre-1967 Israel) with Palestinian settlements located beyond the 'Green Line' (Figure 5.9). The same is true of the Jewish neighbourhoods which have been founded in East Jerusalem, as well as settlements (such as

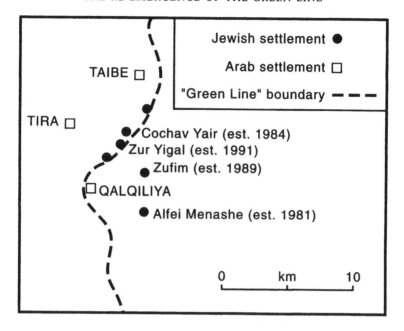

Figure 5.9 Jewish settlements as wedges in the way of territorial exchange

Ma'aleh Adumim and Pisgat Ze'ev) around the periphery of the city. These, in effect, create Jewish residential concentrations in and between the Palestinian neighbourhoods, resulting in ethnic heterogeneity at the regional level and making future partition along ethnic lines increasingly difficult.

Clearly the nature of the political boundary between an Israeli and Palestinian entity will have a significant influence on the potential economic viability of the newly created state. The 'sealing' up of boundaries between an Israeli and Palestinian political entity will result in severe economic dislocation for both states – in terms of the sudden change in spatial orientations of markets for employment and outlets for consumer goods. This is likely to have far greater consequences (certainly in the short term) for the Palestinian entity. The West Bank is blessed with few natural resources, nor does it possess a self-sustaining industrial infrastructure. Since 1948, the West Bank territory has been discriminated against, both by Jordan (prior to 1967) and Israel (since 1967) in terms of capital investment in the local economy. Nor is it clear that, given heavy investment by outside powers following the establishment of an independent state, the West Bank would have sufficient local advantages to make it competitive with the neighbouring economies, particularly that of Israel.

Open borders with Israel offer greater economic advantages (both because of Israel's advanced industrial consumer economy as well as providing the geographical link with European markets) than does a similar arrangement with Jordan. In addition, an open border policy between the new state and Israel would enhance the nature of political relations between the two entities. While 'normalisation' of relations between two such bitter enemies could not be expected to take place in the short term, the economic dependency of one for the other would ensure increased contact and exchange between the two populations. It must be assumed that linkages of this sort can only be beneficial to the political relations between the two entities. The current, albeit asymmetrical, economic interlinkages are to the benefit of both sides, offering cheap labour to Israel and employment opportunities (albeit cheap, menial labour) to the West Bank Palestinians for at least as long as it takes for a vibrant local Palestinian economy to take shape.

Another problematic aspect arising out of the creation of a separate state entity in the West Bank and the Gaza Strip concerns the nature of the territorial link between the two micro-regions. Without free and direct access to the Gaza Strip, the West Bank would find itself in the unhappy position of being a land-locked state, with all the functional implications of such a status (Reitsma 1984). It would be necessary to have some form of territorial corridor linking the two territories, providing customs-free access for goods emanating in the West Bank and being exported through a port in the Gaza Strip. The extent to which such a corridor could be established is largely dependent on the nature of the political relations between the new Palestinian state and Israel. While these concluding comments are no more than scenarios – in the absence of any clear indications of a lasting political solution including some form of territorial re-partition – they come nevertheless to emphasise the importance of the boundary in any future arrangement. Whatever the formal designation of such a boundary, it is clear that the functional characteristics must tend towards contact – at least at the economic level. The case of Western Europe in the post-Second World War era has proved that progressive degrees of economic integration over a long time period can bring about gradual political normalisation. The sealing of boundaries between any future state entities in this region can only harm the long-term processes of political stabilisation.

NOTES

1 The author expresses his gratitude to Peter Lupin of the Department of Geography at Ben Gurion University of the Negev for his preparation of some of the figures, and to Gidon Berman for his comprehensive search of newspaper archival material.

REFERENCES

Alpher, J. and Feldman, S. (1989) *The West Bank and Gaza: Israel's Options for Peace*, Tel Aviv: Jaffee Center for Strategic Studies, Tel Aviv University.
Applebaum, L. and Newman, D. (1991) *The 'Private Sector' Settlements in Israel: Developmental Process and Local Government Status*, Rehovot: Center for Development Studies.
Benvenisti, M. (1984) *The West Bank Data Project: A Survey of Israel's Policies*, Washington: American Enterprise Institute.
—— and Khayat, J. (1988) *The West Bank and Gaza Atlas*, Jerusalem: West Bank Database Project, Jerusalem Post Publications.
Brawer, M. (1984) 'Dissimilarities in the evolution of frontier characteristics along boundaries of differing political and cultural regions', in N. Kliot and S. Waterman (eds), *Pluralism and Political Geography*, London: Croom Helm.
—— (1990) 'The "Green Line": functions and impacts of an Israeli–Arab superimposed boundary', in C. Grundy-Warr (ed.), *International Boundaries and Boundary Conflict Resolution*, Durham: IBRU Press.
Cohen, S.B. (1986) *The Geopolitics of Israel's Border Question*, Tel Aviv: Jaffee Center for Strategic Studies, Tel Aviv University.
Drysdale, A. and Blake, G.H. (1985) *The Middle East and North Africa: A Political Geography*, Oxford: Oxford University Press.
East, W.G. and Prescott, J.R.V. (1975) *Our Fragmented World*, London: Macmillan.
Efrat, E. (1982) 'Spatial patterns of Jewish and Arab settlement in Judea and Samaria', in D. Elazar (ed.), *Judea, Samaria and Gaza: Views on the Present and Future*, Washington: American Enterprise Institute.
Foucher, M. (1987) 'Israel/Palestine which borders? A physical and human geography of the West Bank', in P. Girot and E. Koffman (eds), *International Geopolitical Analysis*, London: Croom Helm.
—— (1990) 'Cross border interactions: realities and representations', in C. Grundy-Warr (ed.), *International Boundaries and Boundary Conflict Resolution*, Durham: IBRU Press.
Harris, W.W. (1980) *Taking Root: Israeli Settlement in the West Bank, Golan Heights, Gaza Strip and Sinai, 1967–1980*, New York: John Wiley.
Kipnis, B. (1991) 'Geographical perspectives on peace alternatives for the Land of Israel, war and peace', in N. Kliot and S. Waterman (eds), *The Political Geography of Conflict and Peace*, London: Bellhaven.
Lynch, K. (1960) *The Image of the City*, Cambridge, Mass.: MIT Press.
Newman, D. (1985) 'The evolution of a political landscape: geographical and territorial implications of Jewish colonization in the West Bank', *Middle Eastern Studies* 21(2), 192–205.

—— (1989) 'Civilian and military presence as strategies of territorial control: the Arab–Israel conflict', *Political Geography Quarterly* 8(3), 215–27.

—— (1990) 'Overcoming the psychological barrier: the role of images in war and peace', in N. Kliot and S. Waterman (eds), *The Political Geography of Conflict and Peace*, London: Bellhaven.

—— (1991) *Population, Settlement and Conflict: Israel and the West Bank*, Cambridge: Cambridge University Press.

—— (1992) 'Colonia in suburbia: reflections on 25 years of Jewish settlement in the West Bank', Paper presented to Israeli–Palestinian Research Seminar on the Peace Process, Rome, February 6–12.

Portugali, J. (1991) 'Jewish settlement in the occupied territories: Israel's settlement structure and the Palestinians', *Political Geography Quarterly* 10(1), 26–53.

Prescott, J.R.V. (1987) *Political Frontiers and Boundaries*, London: Allen & Unwin.

Reichman, S. (1986) 'Policy reduces the world to essentials: a reflection on the Jewish settlement process in the West Bank since 1967', in D. Morley and A. Shachar (eds), *Planning in Turbulence*, Jerusalem: Magness Press.

Reitsma, H.J. (1984) 'Boundaries as barriers – the predicament of land-locked countries', in N. Kliot and S. Waterman (eds), *Pluralism and Political Geography*, London: Croom Helm.

Romann, M. (1989) 'Divided perceptions in a united city: the case of Jerusalem', in F.W. Boal and D.N. Livingstone (eds), *The Behavioural Environment*, London: Routledge.

—— and Weingrod, A. (1991) *Living Together Separately: Arabs and Jews in Contemporary Jerusalem*, Princeton: Princeton University Press.

Semyonov, M. and Lewin-Epstein, N. (1987) *Hewers of Wood and Drawers of Water: Non-citizen Arabs in the Israeli Labour Market*, Cornell University: ILR Press.

Shafir, G. (1984) 'Changing nationalism and Israel's "open frontier" on the West Bank', *Theory and Society* 13, 803–27.

Soja, E. (1970) *The Political Organization of Space*, Association of American Geographers – Special Publication Series.

Turner, F.J. (1962) *The Frontier in American History*, New York: Holt, Rinehart & Winston.

6

RIVER AND LAKE
BOUNDARIES IN ISRAEL

Gideon Biger

INTRODUCTION

The eastern boundary line of what was formerly Palestine (the state of Israel and the Israeli-occupied West Bank) commences at Tel Dan in the north and continues to the Gulf of Eilat (Aqaba) in the south. It is geographically unique in that it passes to a great extent through bodies of water or is adjacent to them. The Jordan and Yarmuq rivers, the Huleh lake, the Sea of Galilee and the Dead Sea are prominent elements in the landscape and they have influenced the delimitation of the boundary. This kind of boundary line gives rise to numerous legal problems and it is with these problems that this discussion deals.

THE LINE BETWEEN PALESTINE AND SYRIA

The boundary line running from Metulla/Tel Dan to the Gulf of Eilat (Aqaba) was established in two separate stages by the British administration in agreement with the sovereign power on the other side of the boundary. The stretch between Metulla and Hamat Gader (Al Hama) is the section of a border line that evolved as a result of negotiations between Britain and France about delimiting the areas of their authority in the Middle East (Biger 1984: 427–42). On 23 December 1920, the two powers came to a preliminary agreement, according to which a section of the boundary line would pass through a body of water, the Sea of Galilee (Lake Tiberias). It was intended that the boundary line reach the village of Tsemakh (Samakh) at the south of the lake, and from there it was to run through the lake until the mouth of wadi (dry river) Massoudiya, following that wadi to Wadi Jerbah and its sources

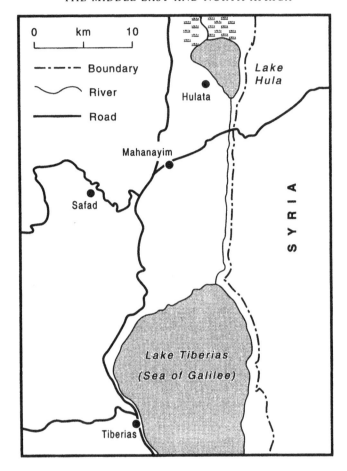

Figure 6.1 The boundary between Huleh (Hula) and the Sea of Galilee

(League of Nations, vol. 22: 355; British Cmd 1195). Such a boundary line would have divided the Sea of Galilee into two unequal parts, giving joint territorial dominion over the two parts of the lake.

The agreement set out general principles concerning the location of the boundary line, but it was the demarcation on the ground that accurately and finally established the boundary line. The boundary line that was finally demarcated, differed greatly from what had been agreed upon in December 1920.

Under the influence of the actual topography as well as in response to political pressures and the desire to interfere as little as possible in everyday life in the border zone, the boundary line was shifted from the location originally agreed upon. In a boundary demarcation agreement initialled on 3 February 1922, and finally ratified a year later on 7 March 1923 (League of Nations Treaty Series, vol. 22: 356; British Cmd 1910), the boundary line between Palestine and Syria was defined as a set of straight lines joining a number of designated points, clearly definable on the ground. Great care was taken in defining the agreement and the description of the boundary line clearly shows that the intention was not to associate it with the River Jordan, the Huleh lake or the Sea of Galilee (Lake Tiberias), but rather to distance it to the east of these water bodies and thus to avoid claims of joint sovereignty over them (see Figure 6.1). Thus, in the section between boundary markers 40 and 42, for example, the boundary line is defined as a straight line to boundary marker 41, on the left bank of the Banias river to about 900 metres south-west of the village of Banias. From there the boundary line continues along the left (eastern) bank of the Banias river in a straight line to boundary marker 42, 700 metres to the north, north-east of Tel Azaziat, along the road running along the left bank of the Banias river. Hence, the Banias river, with the exception of its source at the Banias springs, was completely within the boundaries of Palestine, and there was no joint sovereignty with Syria over it. There is no mention whatsoever of the River Jordan in the description of the boundary line continuing southwards along the Jordan valley, all the boundary markers being to the east of the river. The Huleh lake, too, is completely within the boundaries of Palestine, for boundary markers 49–51 are placed to the east of it. With regard to the section of the boundary in proximity to where the Jordan flows into the Sea of Galilee (Lake Tiberias) the agreement states that after boundary marker 60, the boundary continues parallel to the eastern tributary of the River Jordan and 50 metres to the east of it until it flows into the Sea of Galilee (Lake Tiberias). This section clarifies in detail the intention of the boundary demarcation team to avoid any dependence on the River Jordan as the boundary between Palestine and Syria, the river in its entirety remaining within the boundaries of Palestine.

In the region of the Sea of Galilee (Lake Tiberias) an important legal principle was established. The agreement declares that from the mouth of the Jordan and until boundary marker 61 at the Messifer hot springs, the boundary should run at a distance of 10 metres from the brink of the Sea of Galilee (Lake Tiberias), taking into account possible future fluctuations in the location of the water edge that might occur as a result of a rise in the water level induced by a dam being built on the Jordan south of the Sea of Galilee (Lake Tiberias). The phraseology indicates conclusively that the intent was to include the lake legally within the boundaries of Palestine, in no way giving Syria any territorial rights of access with respect to sovereignty or control over the lake. Apparently, the phenomenon of fluctuation in water level resulting from variation in precipitation was not appreciated, nor was much water pumped out of the lake at the time. However, consideration was given to plans for a hydroelectricity plant using the Sea of Galilee (Lake Tiberias) as its reservoir, and to possible fluctuation in the lake's water level due to dam building at the lake's Jordan outlet. This plant, built later but with modifications to the original plan, was the reason for the addition of a specific clause to the agreement. The effect of the clause was that the Palestinian government or persons designated by that government for the purpose, would have the right to build a dam in order to raise the water levels of the Huleh lake and the Sea of Galilee (Lake Tiberias) above their normal level, on condition that they agreed to pay suitable compensation to owners or occupiers of land that might be flooded as a result. However, the agreement did establish that all usufruct rights of Syrian inhabitants to the waters of the Jordan and the Sea of Galilee (Lake Tiberias) would remain in force. The clause makes it clear that all of the River Jordan (and also the major part of the Banias river) and the entire Sea of Galilee are part of the territorial area of Palestine without shared ownership or dominion with the neighbouring state. Neverthe-less, they do confirm the rights of usufruct (irrigation, fishing and voyage) of those inhabiting the land across the border, conforming with the desire to ease everyday life as much as possible in the border zone.

The establishment of the state of Israel and the 1948–9 war abolished the old boundary lines. The Security Council decision to partition Palestine upheld the boundary line between Palestine and Syria, but the events of the war changed the location of the line. In the armistice agreement with Syria, Israel failed to recover some of the territory that had been part of Palestine during British rule. Areas conquered by the Syrians became 'demilitarised zones', the line from the Huleh lake to the Sea of Galilee (Lake Tiberias) was shifted westwards to the Jordan, and the section of the boundary alongside the Sea of Galilee to a point north of Ein-Geb was switched to the water's edge (Bar-Yaacov 1967) (see Figure 6.2). Syria as a state (not merely the neighbouring Syrian villagers) was thus entitled to claim rights of usufruct regarding the waters both of the Jordan and the Sea of Galilee (Lake Tiberias). These lines were drawn as temporary lines of separation but in effect they continued to be operative in the area until June 1967.

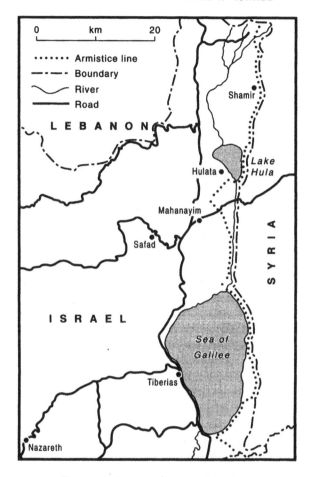

Figure 6.2 Armistice line: Israel–Syria

They resulted in numerous boundary conflicts between Israel and Syria concerning dominion over the Jordan waters and the rights of usufruct in the Sea of Galilee (Lake Tiberias). The war in 1967 re-adjusted this temporary state of affairs; the line is now far away from the Jordan and the Sea of Galilee (Lake Tiberias), to the east. None the less, it may be assumed that any final political settlement will take into consideration the original boundary line in this area and attention will have to be given to its location and its features in the future.

THE EASTERN BOUNDARY LINE:
PALESTINE–TRANSJORDAN

The section of the boundary from Hamat Gader (Al Hama) to the Gulf of Eilat (Aqaba) was determined separately from that described above. The boundary of Palestine with Transjordan was decided in 1922. In effect, it was merely an administrative border between two administrative units under one authority – that of the British High Commissioner for Palestine and Transjordan. The line was initiated (Alsberg 1967: 230–50, Gilhar 1979: 47–69, Biger 1981: 203–6) by order of the High Commissioner without any pact or agreement being signed between the various parties. This boundary line was described as a line starting at a point 2 miles west of Aqaba town on the gulf of that name. The line continues along the middle of Wadi Arabah, the Dead Sea and the River Jordan to the point of its juncture with the Yarmuq river, and thence along the middle of that river until its juncture with the Syrian boundary (*Palestine Gazette*, 1 September 1922) (see Figure 6.3). International recognition of this boundary rested upon a memorandum concerning the amendment of those paragraphs of the mandate that referred to Transjordan. Britain transmitted this amendment to the Permanent Committee for Mandate Affairs at the League of Nations on 23 September 1922, and the Secretary of the League of Nations ratified the status of the boundary line in his answering letter. Reference to the line in the autonomy agreement between the British government and Emir Abdullah in 1928 granted additional legal sanction to the boundary line described above (Great Britain Treaty Series no. 7 1928: Cmd 3488).

This boundary line was not delineated very accurately on maps or demarcated on the ground either at the time of its definition or later. From an exchange of telegrams between the Colonial Office and the High Commissioner for Palestine at the time the boundary was decided, it is apparent that the High Commissioner, Sir Herbert Samuel, intended the boundary to be drawn so as to divide the Dead Sea and the River Jordan accurately into equal parts, because Transjordan, in his opinion, would raise serious objections if the boundary did not run through the middle of the River Jordan (PRO 1922, Toye 1989). In the Yarmuq river, too, the line was to go through the middle of the flow to point no. 124R 32 on map no. 5 of the PEF series. Unlike the decision made with regard to the River Jordan north of the Sea of Galilee (Lake Tiberias), here it was explicitly stated that the intention was to use the River Jordan as the boundary line and not to leave it to the exclusive dominion of one state alone. At the time, the political future of Palestine

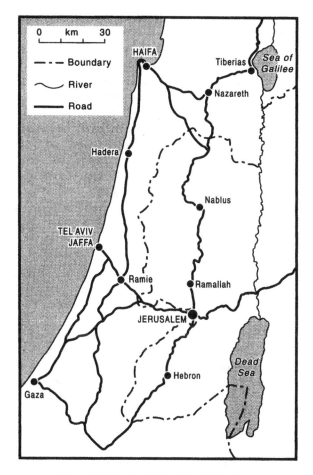

Figure 6.3 Armistice line: Israel–Jordan

and Transjordan was still uncertain. Thus, since the two territorial units were ruled by Britain and the boundary line served little practical purpose, the officials of the British administration were painstaking in defining the boundary line in the river itself rather than on one of its banks. It is conceivable that their intention was to force Palestine and Transjordan to exploit the waters jointly and thus to guarantee peace and cooperation between the Arab state, Transjordan and the future Jewish state of Palestine. The officials were content to leave the final definition of the line as 'the centre of the Dead Sea and the Jordan', although the wide experience of British statesmen might well have encouraged a more accurate definition of the line.[1] In one of his telegrams to Britain, the High Commissioner proposed a line from the Gulf of Aqaba through the middle of the Arabah wadi and the Dead Sea to Beisan, and then along the eastern boundary of the Tiberias sub-district. The British Colonial Office, however, demanded a clearer legal definition that should rely on elements in the landscape or geographical lines. The definition would have to satisfy queries concerning as to which of the Jordan's banks was the boundary, as to whether half of the waters of the Dead Sea were to belong to Palestine, and clarify the exact location of the eastern boundary of the Tiberias sub-district. The Colonial Office was well aware of the inadequacy of boundary definitions based on local, administrative borders, which are liable to change.

Ostensibly, the section of the boundary line from the Dead Sea to the Syrian border, a boundary line running through a river, required no boundary markers. Nevertheless, hydrological events brought about problems regarding the location of the boundary. The flow of the River Jordan was not constant. Rainstorms and consequent torrential increase in the level of the river would bring about changes to its channel. In the winter of 1927/8, the Jordan overflowed and shifted its bed westwards, leaving a strip of land about 800 metres wide between the old and the new river beds. The High Commissioner at the time, Lord Plumer, requested the advice of Colonial Office experts in the matter of the true boundary line, in view of the frequent changes in the flow of the river (Biger 1981: 205). Unable to decide, the Colonial Office requested advice from the Foreign Office, since it was impossible to determine the exact location of the 1922 boundary line. An examination of the difficulties concerning the location of a boundary in a river and study of similar cases in other parts of the world, did not result in an unequivocal answer. It was found that in some cases boundaries seemed to follow wherever the river flowed, irrespective of changes in the channel of its flow. Conversely, in many cases changes in the river channel were not followed by changes in the boundary line, which remained in its location along the original channel of flow existing when the boundary line was defined (PRO 1927). The boundary issues in Palestine were, therefore, left to the High Commissioner's decision on the spot. Following the strong demands of Transjordan to move the boundary line in all cases to the middle of the river, and since it was impossible to keep track of every change in the river's channel of flow, Lord Plumer decided that the boundary line between Palestine and Transjordan would always remain in the middle of the river, and any change in the channel of flow would in fact result in a transfer of territorial sovereignty from one authority to the other. This principle regarding change in the river flow is still valid today, and is applied to shift the 1949 armistice line or the 1967 cease-fire line between Israel and Jordan.[2]

The boundary of Palestine, from Hamat Gader southwards, was delineated in a general way on maps, but no markers were set up on the ground nor was an attempt made to survey the Dead Sea, the Yarmuq or the Jordan in order to determine the exact location of the 'middle' of the water body.

In 1946, following the grant of independence to Transjordan, the British administration found it necessary to define the exact location of the boundary. What had been a local, administrative boundary was now to be an international boundary line between the independent Kingdom of Transjordan and the territory of Palestine, still under British rule. Only a small section of 4 kilometres was demarcated in the region of the Gulf of Aqaba, and the exact delineation of the water boundary line between Palestine and Transjordan was not effected.

The establishment of the State of Israel and the armistice agreements brought into effect material changes in the location of the boundary line. The 'armistice line' existed between 1949 and 1967, and its northern section, from a point south of Tirat Zvi to the junction with the Syrian border, was congruent with the British line along the Jordan and the Yarmuq. In the southern section of the line in the Dead Sea, a line was drawn on the maps from a point north of Ein Gedi to the middle of the sea. From there it was drawn congruent with the British line to the south coast of the sea. Neither the line in the Jordan and Yarmuq nor that in the Dead Sea were the products of accurate survey measurement or the results of a special agreement with the Kingdom of Jordan. In effect they were simply copied from earlier maps. The boundary line south of Tirat Zvi on the Jordan and the northern half of the Dead Sea was unilaterally abolished by the Kingdom of Jordan when it annexed the territory (the West Bank) of Palestine between the western boundary of the Kingdom of Jordan – the River Jordan – and the armistice line with Israel. The effect of this act of annexation was to abrogate all need for a line separating the territory on either side of the Jordan. Although the annexation was not recognised by the world, with the exception of Britain and Pakistan, there was in fact no line of separation in this section between 1949–67. The 'Six Day War' created the 'Cease Fire Lines', marking on maps the position of troop confrontation after fighting had ceased. The 'Cease Fire Line' was delimited east of the 'Armistice Line' and established along the old boundary line. Once again, the line, which went down the middle of the Yarmuq, Jordan and the Dead Sea, was delineated on the maps without any survey measurement on the ground and, in fact, was copied from maps previously prepared during the British mandate. Unlike the armistice agreements of 1949, no boundary agreement was signed concerning the 'Cease Fire Lines', which merely represented a temporary military situation.

It is to be assumed that the Yarmuq or Jordan rivers, or at least some sections of these or other rivers, will be utilised in the delimitation of the boundary in any future agreement between the State of Israel and the Kingdom of Jordan on the subject. It will also be necessary to be very precise when marking the boundary line running through the Dead Sea. There have been very real changes in the shape of this closed lake since 1922 as a result of the use of the Jordan waters for irrigation as well as the work connected with exploiting its mineral resources. These, too, must be taken into account. A study of issues arising out of the delimitation of other boundaries in rivers and lakes in the world may be of assistance in understanding the problems that will arise when the time comes to delimit these boundaries. Recognition of legal precedents can be of help in the accurate and agreed delimitation of the future boundaries of the State of Israel.

SUMMARY

Establishing a boundary in a body of water is only a first stage in a boundary agreement. Description of a boundary line is usually restricted to one or two paragraphs in a boundary agreement, whereas numerous additional clauses deal with the definition of rights of usufruct, ownership, voyage, and irrigation installations with respect to the river waters as well as various matters related to the administration of the border zone. These matters have not been dealt with here, representing as they do topics for separate discussion. None the less, the process of accurately defining a boundary line and, of even greater importance, legal and geographical interpretation of the boundary demarcation team's intentions when marking the boundary, are all matters for detailed examination both by legal experts and by historical geographers (Biger 1988: 341–7). Boundary lines passing through bodies of water raise problems because of the geographical character of the territory through which they extend. Rivers and lakes have been used to mark boundaries for a considerable span of time. In consequence, there has been much legal deliberation over the defining of boundary lines in accordance with the processes of change occurring in the landscape.

None the less, a number of problems remain unsolved. Precise cognisance of the judicial issues which could arise from establishing boundary lines in a body of water such as a river or a lake is bound to facilitate future discussions of the boundaries of the State of Israel.

NOTES

1 Maps made during the British rule showed a line along the middle of the Dead Sea, although this was not based on accurate measurement. In view of the narrow breadth of the Jordan and Yarmuq rivers, the line representing the boundary usually covered all the space on the maps allocated to the rivers. It was only on large-scale maps that the line was drawn down the middle of the river; here, too, this was without accurate measurement.
2 A rise in the level of flow accompanied by a westward shift of the channel near Ashdod Yaakov occurred in the winter of 1978/9. The result was the transfer of a number of *dunams* of land, including a banana grove, from Israel to Jordan.

REFERENCES

Alsberg, A.P. (1967) 'Setting the eastern boundary of Palestine', *Zionism*, 230–50.

Bar-Yaacov, N. (1967) *The Israel–Syrian Armistice: Problems of Implementation*, Oxford: Hebrew University Press.

Biger, G. (1981) 'Settling the eastern boundary of Mandatory Palestine', *Cathedra*, no. 20, 203–6.

—— (1984) 'Geographical and political issues in the delimitation of the northern boundary of Palestine during the Mandate period', in A. Shmueli *et al.* (eds), *The Book of Galilee*, Haifa.

—— (1988) 'Physical geography and law: the case of international river boundaries', *Geojournal* 17 (3), 341–7.

British Cmd 1195, 23 December 1920.

—— 1910, 7 March 1923.

Gilhar, Y. (1979) 'The separation of Trans-Jordan from Palestine', *Cathedra*, no. 12, 47–69.

Great Britain Treaty Series, no. 7, Cmd 3488, 20 February 1928.

League of Nations Treaty Series, vol. 22, 355, 356.

Palestine Gazette, 1 September 1922.

PRO (1922) File 42771, CO/733/24.

—— (1927) Despatch 1284, CO/733/42, File 44550.

Toye, P. (ed.) (1989) *Palestine Boundaries 1833–1947*, Farnham Common: Archive Editions.

7

GAZA VIABILITY
The need for enlargement of its land base
Saul B. Cohen

INTRODUCTION

Since the Gulf War's end, the United States has mounted a vigorous international effort to bring Israel and the Arabs to the negotiation table.[1] There have been other initiatives to seek a resolution to the conflict. This one differs because it is based upon a broad international consensus and because the emphasis is upon the process of negotiations, not the end result. The effort is not a solo American one. The Soviet role was critical in influencing the Syrian stand. Moreover, American insistence upon a European Community and United Nations presence, in spite of Israel's initial opposition, recognised the important stake of Maritime Europe in the Middle East, and the revitalised role of the United Nations as a consensus-seeking mechanism, rather than an instrument of the Cold War.

With respect to the protagonists themselves, surely the earth-shaking events of the past two years – the end of the Cold War, the shattering defeat of Saddam Hussein, the ascendance of Egypt to leadership within the Arab world, the *intifada*, and the rising tide of Soviet Jews' immigrations to Israel – favour peace prospects. Without its Soviet patron, Syria can only seek a resolution of the Golan problem through negotiations. Jordan's economic and political plight, exacerbated by the return of Palestinians from the Gulf and the defeat of its Iraqi ally, has motivated King Hussein to try to return to the centre of the negotiations stage. More than ever beleaguered because of the Palestinian Arabs, the failed political and military strategies of the PLO and the economic consequences of the *intifida*, the protagonists are now desperate for peace. Most have come to the realisation that they have lost the war option. Israel, facing enormous economic problems in absorbing the Soviet immigration, may also be more amenable to compromise. Its military supremacy within the region, and the prospects that the United States

110

will continue to help it maintain that supremacy without having to contend with Soviet countermeasure, make it feasible for Israel to take political risks in the pursuit of peace.

All parties apparently agree that Jerusalem should be left to the last. There the problem is to find a political/administrative structure that makes due provision for national, international and local interests and yet keeps the city open and unified. Boundaries in this case will have to be lines of accommodation, not separation.

There also seems agreement that the West Bank–Judea and Samaria should be treated in stages in accordance with the lapsed aspects of the Camp David agreement. The first period, local autonomy and Palestinian self-rule, would then be followed by negotiations over the ultimate disposition of the territory. At that stage the question of where to draw the boundaries between Arabs and Jews will be the key issue. There are many who feel that the peace breakthrough will most easily be achieved in the Golan Heights. Clearly, Assad has only agreed to come to a conference if he can pursue the 'land for peace' option. Despite Shamir's demurrals, many Israelis suspect that the Likud and the Israeli Right will compromise on the Golan because they have little religious or ideological stake in the region. However, the governing coalition in Israel cannot guarantee any peace agreement with Syria that involves the return of land. Only Labour and the Left can. Within a nearly divided Knesset, the Likud needs all of Labour's votes to offset the large numbers of its own right-wingers within its coalition who are opposed to any territorial compromise.

For Labour, which initiated settlement on the Golan to protect its Galilee communal and cooperative villages – the nuclear core of the Socialist Zionist state – the Golan Heights are viewed as strategic assets. Labour's commitment to territorial compromise is not likely to forgo continued Israeli control of the western edge of the escarpment, the northern, Mount Hermon source of the Banias River and the defensive positions in the south and south-east that overlook the Yarmuk and Raqqad gorges that are defences against Syrian, Iraqi and Jordanian threats. Therefore, the bargaining over the Golan is bound to be hard and protracted.

For Gaza, the issues of boundaries *per se* are not vital. Rather, the question is the size and content of the land base. The area should prove more amenable to negotiations because Egypt, Israel, Syria, Jordan and even the West Bank Arabs have little psychological and strategic stake in Gaza. Indeed, the subtitle for this page might be 'Gaza – a place that only Gazans want'.

WHY GAZA?

The rationale for peace negotiations focusing attention on Gaza is that there is a chance to make tangible progress, especially if pressure is brought to bear by the international community, and external economic help is provided. But only a territorially enlarged Gaza would have the capacity to make Gaza independence a reality, rather than an economic trap.

Israelis are far less militarily concerned about Gaza than they are about the West Bank, and it has no religious symbolism to them as an area. Public opinion polls consistently show that 85 per cent of the Israeli public supports withdrawal from Gaza. It is peripheral to the interests of the adherents of Eretz Yisrael Ha Shlema (the 'Undivided Land of Israel'). Historically, Gaza was not part of the Israelite Kingdom, having been retained by the Philistines during the Israelite conquest of Canaan. The few Jewish settlers there, who number only about 2,500, many of them students in religious schools, live in fourteen villages and campuses, nearly all situated on the sand dunes adjoining the coast at the southern end of the Strip.

Strategically, Gaza holds few problems for Israel. Even in the event of a rupture of Israeli–Egyptian relations, a Gaza that is *not* under Israeli control would be vulnerable to overwhelming Israeli military pressure because it can be so easily cut off and enveloped. The Gaza Strip was a threat to Israeli border settlement security from 1950–6 by actions of the Egyptian Army and the PLA (Palestine Liberation Army), and especially by *fedayeen* action from 1955–6. Again in 1969–72, although under Israeli occupation, it was a centre for terrorist activities against Israel. However, the Israeli Defence Forces are now far better able to contain such activities, given their high-technology surveillance devices and encircling positions.

The *intifada* has put Gaza in perspective for many Israelis. The 'ungovernability' of the Strip and the cost of clamping down on the uprising have moved various Israeli groups to advocate unilateral military withdrawal, hermetically sealing off the Strip and leaving it to its own devices as it seeks some form of limited autonomy.

However, unilateral withdrawal from Gaza as it now is constituted would leave the Strip destitute and in turmoil. In all likelihood it would become a PLO or Hamas (the fundamentalist Islamic movement) base against Israel (JCSS Study Group 1989). Such an eventuality would require rapid re-entry of Israeli forces. A Gaza mini-state sponsored in the first stage by Israel and Egypt is a more viable alternative to unilateral withdrawal.

Past Israeli overtures suggesting possibilities for limited authority for Gaza have come to naught. However, there is now evidence of increasing impatience in Israel with the burdens of the Gaza occupation. Recently, two proposals have surfaced – one from the Israeli Left and one from the Israeli Right, each groping for a solution for Gaza. On 10 December 1990, Labour MK Yossi Beilin, former deputy Foreign Minister, announced on behalf of Labour's Mashov Circle (a liberal, intellectual faction) that his group would seek to amend the Labour Party's platform to advocate a sovereign Palestinian entity in Gaza that would be the core of an eventual independent Palestinian national entity. While this has met with opposition within party ranks, it nevertheless has become a matter of public debate. In February of 1991, Labour Party head Shimon Peres called for unilateral withdrawal from Gaza.

Beilin's proposal (Beilin 1990) is for a 'Gaza First Peace Conference' involving an elected Palestinian delegation under the guidelines agreed upon between Israel and the United States in January 1987 and between Israel and Jordan in the London Agreement of April 1987 between Foreign Minister Peres and King Hussein. Alternative negotiating frameworks that he sets forth are an international conference or Egyptian–Israeli negotiations leading towards a multinational trusteeship under the United Nations and eventual self-determination for Gaza. The Beilin plan sees Gaza as the first stage of an overall peace process which will be tied to political stability and autonomy in the West Bank, and ultimately a Palestinian national entity in confederation with Jordan.

On 23 December, Defence Minister Moshe Arens appointed a committee to study economic development possibilities in the Gaza Strip as a way of separating the territory's economy from Israel's. Arens, a major leader within the Likud party, acted in consonance with a decision by a ministerial committee headed by Prime Minister Yitzhak Shamir to encourage the Israeli military government to try to set up an independent economic infrastructure.

Clearly, these initiatives have different premises – one that the occupation should cease, the other that Israel should maintain sovereignty over the Strip while separating it from Israel proper economically. Separation would reduce the security risk of having tens of thousands of Arabs from the Territories working in the heart of Israel. Ultimately, the two positions could converge because an economically independent, but politically subordinate Gaza would stand out as a colonial possession at a time when the world no longer brooks colonialism, and the pressures for self-rule would become overwhelming.

Israel cannot unilaterally impose either independence or economic self-sufficiency on the Palestinians of Gaza. As they have spurned overtures to relocate Gaza refugee camps to new housing on the basis of rejecting piecemeal solutions that do not take into account independence and the problem of the West Bank as a whole, so are they likely to respond negatively to such new initiatives. This is the diplomatic strength of the Beilin proposal.

Egypt's participation thus becomes critical; its mediation could tip the scales. Egypt's involvement could reassure the Palestinian Arabs that independence for Gaza would not foreclose subsequent talks on Judea and Samaria, thereby making it politically feasible for the Gazans to enter into the discussion.

Egypt can play the ideal interlocutory role because, while it is a geographical neighbour, it has nothing to gain from annexing Gaza. Indeed, it had no interest in doing so when it occupied the Strip from 1949 to 1967. (Ironically, Israel still applies Egyptian law in the Gaza Strip.) Sinai and the Suez Canal, not Gaza, are Egypt's strategic divide with Israel. The focus of Egypt's planning attention is on North Sinai from the Suez Canal to El Arish where there is potential to relocate populations from the overcrowded Nile Delta. It does not want the additional economic burden of Gaza. Egypt also holds some of the resource keys to Gaza's economic development prospects. By its involvement, Cairo would bring economic credibility to plans for creating the new state.

Prospects for bringing Egypt into the process are good. In fact, Sadat once supported the idea of a 'Gaza First' approach and even suggested adding North Sinai territories to the Strip that would include Rafia and Yamit (Quandt 1986: 149). Solving the Gaza problem would be an important step in vindicating Camp David. It would demonstrate that the peace agreement with Israel had not been made at the expense of the Palestinian Arab cause, but, indeed, had served that cause. It would also reinforce Egypt's return to Arab world leadership, a return already strengthened within the anti-Saddam Arab coalition by its prominent role in opposing Iraq's invasion of Kuwait.

Palestinian euphoria over the support that it has received from Saddam Hussein is now but a bitter memory. The Palestinians will have to look elsewhere and this will inevitably draw them back to the Egyptians – the only Arab power capable of leading a new peace initiative. While the Gazans would reject Israel's supervision of local elections to initiate a self-determination process, they are likely to accept joint Egyptian–Israeli sponsorship.

Table 7.1 Projected land requirements for Gaza state

Land requirement	Area (sq. km)
Housing/cities, towns, villages[2]	250
Hotels[3]	15
Industry[4]	10
Commerce[5]	5
Agriculture[6]	200
Roads and other public infrastructure	100
Shoreline, beach and dune protection[7]	100
Parks and reservations	70
Land reserves	250
Total	1,000

BOUNDARIES, LAND AND OTHER RESOURCE NEEDS

We have spoken of the need for a geographically expanded Gaza. A major problem in dealing with the Gaza question is that most solutions that have been considered are too narrowly framed in political and economic terms (Ben-Shahar *et al.* 1989). A geopolitical approach that specifically takes into account the necessary interplay between people and their environment would be more fruitful. To speak of independence or some form of autonomy for Gaza without reference to its land and resource needs and to the ecological fragility of the overcrowded Strip, is to ignore the problem of viability.

The nearly 500,000 refugees (478,000 according to UNRWA counts, 280,000 of whom are crowded into camps) could not be resettled in new towns within the present boundaries of Gaza, unless they were to be relocated within a massive jumble of high-rise apartments. But this would be ecologically disastrous, given the terrain of dunes and shifting sands, the shallow water table and the need for diffused water and sewerage run-off catchment areas.

To become a successful mini-state, one that would serve as a 'gateway' or exchange-type of state, Gaza will need additional land, larger supplies of water than are now available to it and access to energy. The Strip is already too small to sustain its present population, let alone to provide for economic development and population growth. Its 362 sq. km (44 km in length by 8 km in width) house an Arab population of 700,000, or a density of about 2,000 persons/sq. km (5,100/ sq. mile) – among the highest in the world for a non-industrialised society. This population is growing at the rate of 3.4 per cent per annum. Gaza covers only 1.3 per cent of the total land area of the Occupied Territories, even though it holds nearly 40 per cent of their Arab populations.

A rough estimate of the need for additional area is another 650 sq. km, for a total of 1,000 sq. km. Table 7.1 provides the desired areas by categories. The data are driven by population increase requirements of 1

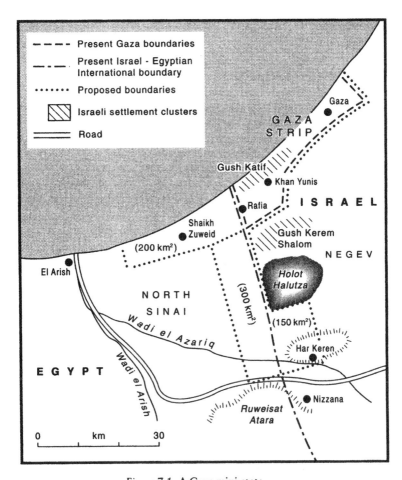

Figure 7.1 A Gaza mini-state

116

million in ten years and 1.2 million in twenty years, and by ecological considerations which call for setting aside land reserves to protect the fragile desert environment, and permit no building within 200 to 400 m from the shoreline. Building calculations are based upon low-rise residences, hotels, factories and commercial enterprises. Clearly, less land area would be required with high-rise building, but ecologically this would be highly undesirable. A series of land requirement options could be developed under alternative building height scenarios.

Additional lands for a Gaza state can only come from Egypt and Israel. The two countries have areas adjoining Gaza that are nearly empty and barren which they can afford to relinquish in the interests of the benefits that they would both derive from resolution of the Gaza question (Figure 7.1).

Ideally, Gaza's boundaries should be extended westward for 30 km along Egypt's North Sinai coast, to a point east of El Arish, and to unite all of Rafia and include the area from which Israel withdrew in accordance with the Camp David agreement. This would provide the new mini-state with an uninterrupted coastal strip of about 75 km. The beach has a gradual slope, and the coastline, which adjoins a 15–20 km-wide underwater shelf is very gently curved – an idyllic sea border. Shifting sand dunes back up against the beach, south of which is a plain of sands, sandstone hills, gravel and alluvium. East and south of the coastal zone are slightly higher lands (150–300 m above sea level). These are largely covered with sand but in some cases, between the dunes, there are pockets of agriculturally suitable fertile loess soils (or sand-covered loess) that are both Aeolian and riverine in nature.

The total area of approximately 1,000 sq. km could be created with the addition of land contributions from Israel and Egypt. The Israeli inland supplement might consist of about 150 sq. km (15 km by 10 km) of empty Negev lands bordering North Sinai. The area would not adjoin the Gaza Strip directly, but would extend from the southern edge of Holot Halutza (the sands of the Halutza syncline) for 15 km to include the sands of Holot Agur and then follow the ridge line of Har Keren (the Keren uplands, 10 km north of the ancient Nabatean site and modern border post of Nizzana), which is approximately 350 m above sea level. The area would be connected to Gaza by inland Egyptian territory that follows the current international border.

The Israeli lands could be matched by the Egyptian contribution of 200 sq. km (30 km by 7 km) from the north-easternmost Sinai coastal strip, and 300 sq. km (37 km by 8 km) of the bordering North Sinai plain, extending southward to the 200 m elevation rise (Ruweisat

Atara) to include Wadi el Azariq and connect up with the lands trans-ferred by Israel. The interior North Sinai plain is very sparsely populated by Bedouins and the Israeli lands are empty. The proposed section of the North Sinai coast contains the western edges of the town of Rafia, which were split away from the rest of the town by the Egyptian–Israeli peace treaty, and the small town of Sheikh Suweid, 10 km to the west of Rafia. The Sheikh Suweid–Rafia area is a flood plain suitable for cultivation as well as land development in general.

The territory of an enlarged Gaza would have a modified T-shape, with the top of the 'T' following the Mediterranean, and the vertical leg straddling the present Israel–Sinai border. Such an area could relieve Gaza's overcrowding, provide for agricultural and natural land reserves, and spread urban activities (including small towns and hotels) to provide a unique, low-rise cultural landscape.

The new boundary line would have to be drawn so that Israel would retain the Kerem Shalom cluster of settlements within its national borders. There, at the junction of the north-west Negev, Gaza and North Sinai, 17 Israeli villages have been established, including some which were displaced from North Sinai, to seal off the southern end of the Gaza Strip. No Israeli political party would agree to dismantle Israeli settlements that are located inside Israeli territory. There are still bitter memories of the 1982 Israeli withdrawal from North Sinai when 18 settlements, including the town of Yamit, were dismantled. The boundary would also be drawn north of the Nizzana–Abu Aweigla road so that Israel's Beer Sheba region would have direct connection to Egypt's El Arish and the coastal road Qantar on the Suez Canal, or the inland highway to Ismailiya on the Canal.

Merging Gaza with the adjoining section of North Sinai would allow for the reunification of Rafia, where a barbed wire fence now marks the border and cuts off Palestinians on the Egyptian side from their services and markets in Gaza Rafia. Elimination of the border at this northern end would resolve the inequities of the boundary that currently follows the 1906 international line between Egypt and Ottoman Palestine, and which takes no regard for the needs of a people living on both sides to be able to interact freely (Brawer 1988: 170–1, 214–15). Placing the border east of El Arish would have historical precedent, for in the nine-teenth century the boundary between Egypt and Palestine extended from El Arish to Suez, at the head of the Gulf of Suez. While Egypt has announced recreational development plans for all of the North Sinai Coast, its efforts are likely to be concentrated in the area of El Arish, the largest urban centre, and the lands extending westward to the Bardawil

Lagoon and the Suez Canal. Only very modest activities are to be found at Sheikh Suweid, and the economic benefits that Egypt can gain from developmental activities in Gaza as well as the political dividends from a genuine peace are greater than the costs of giving up some land.

The future of Israeli farm settlements and religious school campuses located in Gush Qatif (the Qatif cluster) in the south-western end of the Gaza Strip, would have to be addressed. In the long run they should be removed. Israel is not likely to agree to their dismantling. However, they could remain in place through a political accommodation that accords their populations permanent resident rights in the Gaza state, while retaining their Israeli citizenship and special communal status. Just as Israelis might opt to remain in Gaza under these circumstances, so might some Israeli Arabs adopt Gaza or West Bank citizenship at some future time, while maintaining inalienable residential rights in Israel.

Alternatively, the area in which the Gaza Israeli settlements are located could be affixed to a multinational corridor that would extend from the Gaza state across Israeli territory to the West Bank and to Jordan (Cohen 1986: 80–3, 112).

In addition to its land needs, the new state would require access to water and energy. Egypt seems to be in a position to provide these resources. Gaza is a water-deficient region, semi-arid in the north and arid in the south. Its water resources are overexploited (Schwarcz 1982: 95–100). Average rainfall is 295 mm (10 inches) per annum, higher in the semi-arid north (370 mm) where some winter crops can be success-fully grown, and lower in the arid south (220 mm) where dry farming is very marginal, and winter crops are low grade. In general, there is a high dependency on irrigation, especially because the mean annual evaporation is three times the natural rainfall. Citrus, which had doubled in acreage to 18,000 between 1967 and 1980, is by far the largest crop. However, in the past decade acreage has shrunk by over 25 per cent due to water shortages and increase in water salinity. Treefruit crops and livestock acreage are stable. During the same period, field crops have dropped, while vegetable, strawberry and melon production has increased rapidly to 4,000 acres, much under greenhouses at the south-west corner of the Strip. The vegetables are grown intensively along the desert coast under the 'MWASSI' system whereby 2 to 3 m ditches are dug between rows of sand dunes with mechanical diggers. The bottom of the ditch is then covered with a thin layer of silt brought from nearby wadis and mixed with chemical fertilisers. The crops are lightly irrigated by sprinklers because their roots are so close to the groundwater table, 2–3 m below (Brawer 1990).

In spite of the water deficiencies, half of all land in Gaza is agricultural (45,000 acres), two-thirds of this under irrigation (30,000 irrigated acres vs 15,000 acres of rain-fed crops). There are no surface water resources. Instead, wells provide the irrigation from groundwater which occurs in sand and stone aquifers 10 to 50 m below the surface that originate in adjoining Israeli territory. (Rooftop cisterns are used to catch rainwater for some of the household consumption needs.)

While the groundwater, which is replenished by winter rainfall, can be better exploited through increased efficiencies of irrigation, it is necessary to guard against lowering the water table to a point where seawater would intrude into the Gaza aquifers. Already half of the subterranean waters have a salinity greater than 800 mg/litre. As a consequence, future economic development of the Gaza Strip needs to focus on urban pursuits that are less heavy users of water than agriculture. While not the largest employer, farming is currently the main local source of the economy, representing 26 per cent of total income (Kahan 1987).

Egypt's role in helping Gaza to develop is critical for, in the long run, the Gaza Strip might have to import fresh water from Egypt whose Nile now loses several billion cubic metres of water per year which flow out to the Mediterranean. Egypt's Sinai Canal that takes water from the Nile through the North Sinai to El Arish, can be readily widened and extended to the Gaza Strip. There is some concern that diverting additional Nile waters could create the problem of reducing the scouring power of Nile distributaries as they enter the Mediterranean, thus clogging the Delta. An alternative or supplement to the use of Nile waters would be large-scale desalination plants. The cost of waters from the sea is so high that they would have to be used for urban purposes only.

Despite the wide extent of agricultural land, the Gaza Strip is already highly urbanised (over 80 per cent of the population including those in the refugee camps), many of the farm workers living in the towns or refugee camps. In fact, 40 per cent of the population lives in three urban centres: Gaza (population 135,000), Khan Yunis (population 60,000) and Rafia (population 50,000). However, Gaza has few significant industrial enterprises (Benvenisti and Khayat 1988). Instead, the main sources of income are the 60,000 Gazans who have been working in Israel, cash support from UNRWA to residents of the refugee camps, jobs held by UNRWA and other relief agency employees, and cash remittances from Gazans working abroad. The industries that exist are either small-scale crafts or agricultural and fish-processing enterprises. It is important that agriculture, the heaviest of all water users, be downsized in its land and water requirements, and that manufacturing and services replace lands now under cultivation.

Egypt's role in providing energy can also be critical. Egypt has a surplus of natural gas. To meet Gaza's energy needs for future urbanised activities, natural gas can be shipped from Nile Delta gas fields to the Gaza Strip at economically justified prices. A pipeline of about 250 km could be easily constructed (as well as extended later to Beer Sheba) (Ben-Shahar et al. 1989). The export of water and natural gas to Gaza would not only be a source of foreign trade earnings to Egypt, the pipelines would also be of benefit to the development of the Egyptian North Sinai coast.[8]

ECONOMIC PROSPECTS

The Gaza state's most important economic potential lies in tourism. Its climate is ideal and it can become a new Mediterranean Riviera. Unlike Israel's tourist facilities and most of those of Egypt, or such a luxury resort as Cancun, Gaza's should not consist essentially of tightly packed high-rise, luxury hotels. Rather, the 75 km of coast would allow for dispersed leisure vacation centres – perhaps in clusters of 50 to 100, to 250 to 400 rooms, protecting against overuse of fragile ecosystems. In addition, ranch-style facilities could be built in the interior desert to take advantage of speciality tourism oriented to nature and exploration. Our reference is therefore to a mixed tourist centre of hotels, apartments, villas, bungalows and camps, with both large-scale foreign and local investment and management.

As a labour-intensive industry, tourism has the potential to employ up to 0.5 persons for every room. Thus, an expanded Gaza mini-state, eventually with 20,000 hotel rooms, would directly employ up to 10,000 persons in tourism. Using a multiplier value of one for the regional economic impact brings the total up to 20,000 jobs. The broad mixture of facilities, from popular-price to expensive to low-cost hostels, could attract year-round foreign tourist flow, as well as Israeli and Egyptian visitors, and exceed the 60 per cent capacity calculations used in Table 7.1. The luxury centres could develop gambling and entertainment to rival what Beirut formerly offered, and the coastal waters could offer sports fishing. A long-term target of 750,000 annual visitors (half of what Israel now attracts and about 40 per cent of Egypt's foreign tourists) to support 20,000 hotel rooms seems achievable. This is on the scale of the large state-owned holiday resorts developed along the Black Sea littoral at Mamia in Romania and Zlatni Pjasac in Oulparia. An average of 6 person days per visitor, and a 60 per cent occupancy rate, would yield about 2.75 million person days or from $400 million to $700 million per annum (at an expenditure range of $150 to $250 per day). Many tourists are likely to be individuals who link their stays to visits in Egypt and/or Israel.

The climate of the Gaza mini-state is ideal for tourism, with coastal summer temperatures averaging 80°F (26.4°C) in August, the warmest month, and 57°F (13.7°C) in January, the coldest. The summer is perceptibly cooler and drier than that of the coast from Tel Aviv to Lebanon. In the interior, summer night-time temperatures are pleasant (60°F), although winter nights can be cold (47°F). Limited rainfall, minimal cloud cover and western (Mediterranean) breezes enhance the ambience. As a measure of comparison Eilat summer temperatures, during the period that it receives the bulk of its tourists, are much hotter than those of Gaza. Its winter temperatures are about the same or marginally warmer, but exposed to cool winter winds that sometimes sweep down the Arava from the north and north-east.

A balanced and well-planned tourist industry could make the Gaza mini-state the unrivalled Riviera of the Eastern Mediterranean. The combination of unspoiled beach, offshore waters, dunes, desert interior, a warm and dry climate and accessibility to the outside world is unique. But to make it all work, interior land reserves have to be added to the coastal strip. This in turn depends upon the land contributions from Israel and Egypt. Such contributions would not only fulfil humanitarian and economic needs, they would also address the national self-interests of both countries. As partners in a Gaza Authority charged with paving the way for the new state, Egypt and Israel could also plan to extend joint tourist projects within the mini-state to their respective national territories. For Israel, especially, Gaza would represent formidable tourist competition for Eilat, and also for Egypt's developing East Sinai coast which starts at Taba, and for its plans for the North Sinai coast. Jordan's Aqaba resort would also be affected. However, there is a certain measure of winter-time complementarity that would give an edge to the Gulf of Aqaba region, and Eilat's tourism competes for land and environmental quality with the port. In peace, this port and associated industries could be expanded to take up the loss in tourism.

An expanded Gaza mini-state would be able to house its resident populations in new garden-type towns. The first candidates for new housing are those living in the refugee camps. Many of the towns should be located away from the coast in the inland desert plains areas to provide adequate space for expansion of individual housing units, and to guard against overdevelopment of the shoreline and adjoining sand dunes.

Another major economic activity could centre around the frequently proposed new deep-water port to be constructed at Gaza, serving the new mini-state, the West Bank and Jordan. A port project would require political arrangements between Jordan and Israel that would provide for customs-free transit. In fact, Gaza could be a 'Free Port' in the Free Trade Zone tradition of the 'Free Polis' that it was during Pompeii's times. However, unless the port were also able to serve Israel's Beer Sheba region, its economic prospects are dubious. Ultimately, the port and areas nearby could be the western terminus of a duty-free multi-national corridor containing a highway as first envisaged in the Alon Plan (Cohen 1972). Ultimately, such a corridor might include light assembly plants, and possibly the proposed Med–Dead Canal which would integrate Gaza, Israel, the West Bank and Jordan.

With a mixture of employment in tourism, port activities, financial services, education (Gaza could support a higher education institution of 10,000 students), agricultural processing, farming and free-trade zone for activities like assembly of components, Gaza could become a balanced mini-state. Brawer estimates that with the help of Nile water, agriculture in the Gaza Strip can support 30,000 families (180,000 people) through the conversion of farm acreage to the 'MWASSI' system whereby a family can make a reasonable living from 1½ acres (Brawer 1990a). At peace with Israel, its economy could continue to benefit from wages earned by those holding jobs in Israel, although the numbers would be far fewer than the current figures. Financially, the costs of setting up such a mini-state can be readily borne by outside grants and loans, as well as by the proceeds from commercial investments. Constructing 20,000 hotel rooms and related services, 50,000 modest houses for the current refugee camp residents, and a port and road and pipeline infrastructure might cost $4.5 to $5 billion – a modest price indeed for a giant step towards peace.

THE PROSPECTS FOR A SOLUTION

The prospects for a solution in Gaza hinge on an Israeli–Egyptian agreement to promote the new state's economic development, to add land to it, and to guarantee its external security. Palestinian Arab suspicion of Israeli intentions could be allayed by a process whose first phase would be the establishment of a joint Israeli–Egyptian Authority to maintain external security and mobilise the necessary economic resources as a new, demilitarised mini-state emerges. This would be followed by full independence. Negotiations over Gaza can break the stalemate that has dogged Phase Two of the 1978 Camp David agreement, and be a prelude to the broader peace.

The time to address the Gaza problem is now, and the initiative lies with Israel and Egypt. The two nations have shown that they can work with each other, in spite of the various strains and pressures which have marked their relationships since the peace agreement. Their military forces are not arrayed along their boundary; the Taba dispute was arbitrated in 1988; and the Egyptians have been earnest in their commitment to try to safeguard Israeli tourists from terrorist attacks upon them in Sinai and Egypt.

It is important for Israelis to take a realistic view of the future. It's not decades away; it's at hand. Some see risks for Israel in giving up any territory, but creative initiatives are required to gain the rewards of peace. In the development of a Gaza plan, the stages could start with the present land area of the Strip, then include the Egyptian North Sinai area and finally Israel's Negev land contribution. Giving up Israeli territory, even though empty, would represent a major concession for Israel because of concerns that it would set a precedent for later negotiations with the Arabs. On the other hand, Israel could cite its ceding of Negev lands as a quid pro quo for holding on to certain tactically important West Bank areas when negotiations get under way there. Moreover, Egypt would not allocate land if Israel were to refuse to do so, and giving up land to Gaza would be equally difficult for it.

For Israel, the political problems of not dealing with Gaza immediately are especially acute. Religious fundamentalism has become increasingly entrenched, as the Hamas movement has gained in strength vis-à-vis the PLO. Israel's recalcitrance in dealing with the PLO even indirectly reinforces the position of Islamic extremists who represent an even greater obstacle to accommodation than the secular radical PLO.

Allied victory in the Persian Gulf, the end of the Cold War, the erosion of the PLO's financial support because of its tilt towards Iraq, the importance of the Arab states aligned in the Allied coalition and Israel's inability to play a strategic role in the war because of America's dependence on its Arab partners – these are all part of the new Middle Eastern geopolitical reality. They reflect a new political reality which Israel can disregard only at its long-term peril.

Gaza's poverty and overcrowding cry out for immediate relief, a condition that cannot be met under the present political circumstances.

A bold initiative to bring peace to this area could attract world-wide economic support on humanitarian as well as political grounds. It would also be the start of a broader accommodation on the West Bank, in the Golan, in East Jerusalem and with the Arab states.

There can be no economic development for Gaza without self-determination, and self-determination without economic development is a snare and delusion. Neither development nor self-determination are possible without a coordinated Israeli–Egyptian initiative, and especially without substantial territorial enlargement of the Strip.

This chapter presents a vision and a plan for a modern gateway state – the Riviera of the Eastern Mediterranean. It is very possible, however, that what has been proposed would be culturally unfeasible even if Gaza should attain independence. A state controlled by Hamas is most likely to reject the concept of a hotel/recreational-based economy that would offer the same types of services available elsewhere in Mediterranean tourist centres. Many of the ingredients needed for a large, pulsating hotel and travel industry – from bathing briefs to mixed swimming, to a night-life of dancing, gambling and liquor – are unacceptable to Islamic fundamentalists. The fact that one Arab land, Lebanon, once housed such a thriving industry in Beirut and other sea and ski resorts is only a limited guide to Gaza's future. For the hedonism of Lebanon's tourist life flourished within a Christian-dominated government and society.

Without a massive recreation industry to support 20,000 jobs directly and perhaps another 40,000 indirectly, Gazans would continue to be dependent on work in Israel. The domestic economy would lose the direct benefit of $400 million to $700 million per annum, for which job remittances by those working in Israel would compensate by only about $200 million to $250 million. These calculations exclude contribution of the recreation industry to construction, furniture manufacturing, horticulture, and land, sea and air transportation. While the demand for overall land area would be only marginally decreased, perhaps from 5 to 10 per cent, the political economic nature of Gaza would be radically different from what is envisaged in this chapter and the state's economic prospects would be considerably dimmer.

All too often we deal with boundary delimitation in territorial disputes as ends unto themselves. In the case of Gaza, the boundary issue is only a means. The boundary is an envelope and what is important is the size and content of that envelope. If it seems unrealistic and utopian to bring forth suggestions for the peace negotiations that ask Israel and Egypt to provide 'gifts' of land not under debate, I must content myself with the observation that all must pay a price for a genuine peace. In international politics the line between dreams and 'harsh realities' is often breached – and sometimes in most surprising ways. We've seen this in the world-shaking events of the past two years. Our hopes are to see it again in the resolution of the Arab–Israeli conflict. Gaza can be the first chapter in a new book of peace.

NOTES

1 This chapter was written in the summer of 1991, prior to the historic Madrid conference. Since that time the 'Peace Process' has survived numerous rounds of negotiations without breaking down completely, yet has yielded no significant agreements. Thus the options for alternative conflict resolution outlined in this chapter are perhaps even more relevant now than at the time of writing.
2 Includes 150 sq. km for 25 resettlement towns. Present built-up area is 60 sq. km, with population density of 10,000 sq. km. Proposed urban density is 2,500/sq. km. Figures are based upon low-density housing of 1.5 acre units and urban population of 1 million in 20 years.
3 Estimated 20,000 rooms and facilities. Total floor area of 15 million sq. km × 6 for attached parking, gardens and sports facilities.
4 Estimated 50,000 workers at 30 sq. m floor area/worker and 90 sq. m/worker building surround. Based on 20 per cent of workforce of 250,000 (half of population of 500,000, ages 16 and over, from total population of 1.2 million).
5 Estimated at 6 sq. m floor area/person and 18 sq. m for building surround.
6 Based on 30,000 farm families at 1.5 acres/farm, and 1,000 families at 10 acres/farm.
7 No building 200 m to 400 m from shoreline.
8 This chapter has not dealt with the question of the offshore territorial arrangements that would have to be made in the event that a Gaza, and/or West Bank/Gaza state were to emerge. Gerald Blake (1988) suggested that the waters and seabed of offshore Palestine would be a great asset to the land-locked West Bank as well as to Gaza itself. His map of a 12-mile territorial sea and a conically shaped Exclusive Economic Zone 200 nautical miles from the coast, raises questions not only of the sea's exploitation potential, but of how environmental and navigation controls can be cooperatively developed that will accommodate the needs of the abutting Egyptian and Israeli sovereignties. A detailed study of Gaza's potential should take into account the offshore territory and the problems attendant upon delimiting the sea boundaries as well as those on land.

REFERENCES

Beilin, Y. (1990, undated), 'Self-determination in stages', Internal memo, Jerusalem.
Ben-Shahar, H., Fishelson, G. and Hirsch, S. (ed. M. Merhav) (1989) *Economic Cooperation and Middle East Peace*, London: Weidenfeld & Nicolson.

Benvenisti, M. and Khayat, S. (1988) *The West Bank and Gaza Atlas*, Jerusalem: West Bank Database Project.

Blake, G.H. (1988) 'Offshore Palestine?', *Durham University Geographical Society Journal*.

Brawer, M. (1988) *Israel's Boundaries – Past, Present and Future* (in Hebrew), Tel Aviv:Yavneh Publishing House.

—— (1990a) 'Geographical and other arguments in delimitation in the boundaries of British Palestine', in C.E.R. Grundy-Warr (ed.) *International Boundries and Boundary Conflict Resolution*, Durham: IBRU Press.

—— (1990b) Private memo.

Cohen, S. (1986) *The Geopolitics of Israel's Border Question*, Boulder, Colo.: Westview Press.

Cohen, Y. (1972) *Tacknit Alon (The Alon Plan)* (in Hebrew), Tel Aviv: Hakibbatz Ha Meuchad.

JCSS Study Group (1989) *The West Bank and Gaza, Israel's Options for Peace*, Tel Aviv: Jaffee Center for Strategic Studies, Tel Aviv University.

Kahan, D. (1987) *Agriculture and Water Resources in the West Bank*, Jerusalem: West Bank Database Project.

Quandt, W. (1986) *Camp David: Peacemaking and Politics*, Washington, DC: Brookings Institution.

Schwarcz, J. (1982) 'Water resources in Judea, Samaria and Gaza', in D. Elaza (ed.), *Judea, Samaria and Gaza*, Washington: American Enterprise Institute for Public Affairs Research.

8

THE EVOLUTION OF THE TRANSJORDAN–IRAQ BOUNDARY, 1915–40

Vartan Amadouny

INTRODUCTION

This chapter traces the evolution of the Transjordan–Iraq boundary (now the Jordan–Iraq boundary) from Britain's earliest wartime plans in 1915 for the partition of the Ottoman Empire, through the boundary agreement of 1932, to the modification of the boundary in 1940.[1]

The three principal issues governing the settlement of the boundary were (a) Britain's determination to control the imperial route to the East via the eastern Mediterranean to the Persian Gulf, (b) the challenge to this route apparently posed by Abdul Aziz ibn Saud and (c) the confusion over the exact location of the southern end of the boundary at Jebel Anaza which in turn ultimately threatened to delay Iraq's admission to the League of Nations.

DESCRIPTION OF THE BOUNDARY ZONE

The northern terminal of the Transjordan/Jordan–Iraq boundary as it was defined until 1984[2] lies at the intersection point between the boundaries of Transjordan, Iraq and Syria at Jebel Khurgi at 33°22'29" North and 38°47'33" East, just south of Jebel Tenf. From there it follows a south-southeasterly course to its southern terminal, which marks the triangulation point between Transjordan/Jordan, Iraq and Saudi Arabia at Jebel Anaza, at 32°13'51" North and 39°18'09" East. Most of the boundary zone can be characterised as limestone desert, with the eastern edge of the lava belt in Jordan protruding into the area at the northern terminal. In other areas, wells and grazing grounds traditionally sustained bedouin nomads with customary claims to the territory.

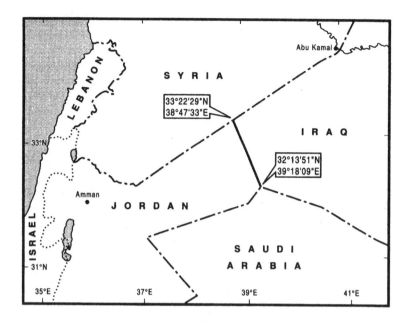

Figure 8.1 The Iraq–Jordan boundary

Near to its midway point, the boundary zone is bisected by the Amman–Baghdad Highway. This was also the route taken in the 1930s by the Iraqi Petroleum Company's oil pipeline from Kirkuk to Haifa, and later by the Haifa–Baghdad pipeline. This pipeline ceased to function in 1948 (see Figure 8.1).

HISTORICAL EVOLUTION OF THE BOUNDARY

The Transjordan–Iraq boundary evolved over a period of 25 years from 1915 to 1940 (Abidi 1965: 8–9, Bureau of Intelligence and Research 1970, Luke and Keith-Roach 1934: 434, Schechtmann 1937: 21–2, and Shwadran 1959: 198–9).[3] Its beginnings lie in Britain's involvement in the First World War in the Middle East, since this established a British claim to certain areas of the Ottoman Empire, while France, Russia, Italy and the Arabs also had claims (Yapp 1987: 275–88).

Britain's war aims in the Middle East were clarified in 1915 by the report of the de Bunsen Committee into 'British Desiderata in Turkey-in-Asia' which established a precedent for the Anglo-French partition of the Ottoman Empire which followed (CAB 27/1, Klieman 1970: 4–9).

129

This clarification was taken two stages further by the 1915–16 agreement with the Hashemites of the Hejaz to grant Arab independence (usually referred to as the Husain–McMahon correspondence) in exchange for their help in a military campaign against Ottoman forces; and by the Sykes–Picot Agreement of May 1916 whereby Britain and France, acting ostensibly to guarantee Arab independence from Turkish rule, effectively arranged to partition the Ottoman Empire between themselves (Moore 1974: 6–21, 25–8).

Britain's long-term objective was to secure the historic routes to India which passed overland from the eastern Mediterranean to the Persian Gulf. The development of air power after 1918 offered a relatively inexpensive military service well-suited to desert conditions. This increased the strategic importance of the largely desert area between Iraq and southern Palestine (Towle 1989: 13–24).

The implementation of these plans was later facilitated by the League of Nations mandates system. At the San Remo Conference of April 1920 Britain and France, regardless of Arab opinion, divided up Syria, Lebanon, Palestine and Iraq into French and British mandates. This partition largely followed the territorial division laid down in the Sykes–Picot Agreement, except that the whole of the former *vilayet* of Mosul came under British jurisdiction.

The French occupied Syria in July 1920, expelling Faisal, the Hashemite Commander of the Arab Revolt from Damascus. By March 1921, at the Cairo Conference, Britain was claiming to have fulfilled its promises to the Arabs by agreeing that Faisal should become the King of the newly created state of Iraq, and his elder brother Abdullah, the Emir of the newly created Transjordan.

Thus, in one manoeuvre, Britain consolidated its hold over the mandates of Iraq and Transjordan, pacified its wartime allies, the Hashemites, and secured the land and air route to India with pro-British governments in Egypt, Palestine, Transjordan, Iraq and Iran.

The one area of instability which remained beyond Britain's control was the heartland of Arabia, where the emerging power of Abdul Aziz ibn Saud threatened the Hashemites' hold over the Hejaz and the integration of Iraq. Britain's attempt to contain the ambitions of Ibn Saud between 1922 and 1925 was part of the process of boundary making by which it aimed to guarantee the security of the Imperial route to the East and to provide the mandated territories with boundaries valid in international law.

Formal boundary making began with the establishment of a boundary between the French and British mandates. The Franco-British

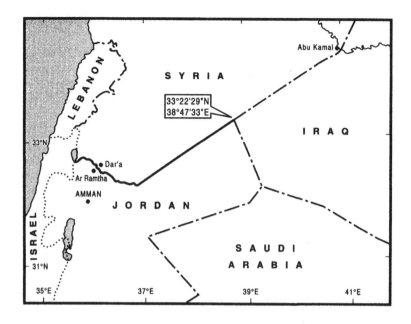

Figure 8.2 The Jordan–Syria boundary

Convention of 23 December 1920 referred to the zone separating Syria from the area to the south (that would eventually become Transjordan), as a straight line running from Abu Kamal in the east to Imtar in the Jebel Druze, continuing in this westerly direction until meeting the boundary with Palestine. This became known as the 'Convention line' (Cmd 1195). The northern terminal of the Transjordan–Iraq boundary was fixed on this line in 1932 by the Syria–Transjordan Boundary Commission. The terminal point was fixed at Jebel Khurgi south of Jebel Tenf. This concurred with the wording of the Transjordan–Iraq boundary agreement which referred to the terminal as 'the nearest point on the Syria–Transjordan frontier to Jebel Tenf'. This was also a slight modification of the Anglo-French 1920 Convention. Where the Convention referred to a straight line running from Abu Kamal to Imtar in the Jebel Druze, the actual location of Jebel Khurgi meant that the straight line ran south of Imtar and just north of Tel Rumah in Transjordan[4] (see Figure 8.2).

Fixing the southern terminal of the Transjordan–Iraq boundary was more complicated because of Britain's strategic need to contain Ibn Saud in Arabia, and because of the latter's desire for a boundary contiguous with Syria. Ibn Saud perceived the Hashemites as a threat to his ambition to unite the Arab tribes under his command, and to recreate the first Wahhabi empire of the late eighteenth and early nineteenth centuries (Troeller 1976: 179–80). He was also unhappy that any proposal to create boundaries between his base in the Najd, and Transjordan as well as Iraq would cut into areas where pastoral nomadism and trading had been conducted for centuries. He therefore wanted a boundary for his territory contiguous with Syria to maintain the time-honoured trading links between Najdi and Syrian merchants.

In addition to this, Ibn Saud favoured a boundary based on territorial advantage as well as human movements. The tribal solidarity, or *asabiya*, of the expanding Wahhabi polity viewed tribes outside the system as ripe for subjugation, their territory and resources included (Al-Azmeh 1986: 76). Thus, Ibn Saud's claim for a boundary with Syria was aimed at separating the two Hashemite states of Iraq and the Hejaz, and giving Ibn Saud territorial advantage to the north of those Arabian tribes which had yet to be conquered (Kostiner 1990: 235).

An indication of where this line might run was provided by Colonel Lawrence in January 1922. Lawrence argued that if Britain wanted to control the imperial route to the east, it should extend Transjordan to encompass the Wadi Sirhan as far south as al-Jauf, an important trading and grazing area with numerous wells. Lawrence envisaged bringing the tribes there under British control, making it easier to contain Ibn Saud in the Najd.[5]

Events in 1922 led Britain in a different direction. Not long after Lawrence had submitted his memorandum Ibn Saud occupied al-Jauf, the base of his main rival, Ibn Rashid. This was followed in March by an attack into Iraq mounted by Faisal al-Duwish and the 'Ikhwan' warriors. This crisis was temporarily resolved by the Treaty of al-Muhammerah of 5 May 1922, one feature of which was an agreement by Ibn Saud to allocate along rather vague tribal lines the border between Iraq and the Najd.

The Ikhwan mounted another raid into Transjordan in August 1922. This was repulsed by the RAF with the Ikhwan sustaining over 500 fatalities, and British and Transjordan forces occupied Kaf in the Wadi Sirhan. Forced back to the negotiating table, Ibn Saud and Sir Percy Cox, High Commissioner of Iraq, agreed an appendix to the Treaty of al-Muhammerah, the Uqair Protocols of 2 December 1922 which defined the emergent Najdi state's boundaries with Iraq and Kuwait (Troeller 1976: 174–5, Collins 1969: 34).

While Britain did not press Transjordan's claims to al-Jauf in these negotiations, the key part of the Uqair Protocols which was to have an impact upon subsequent events was contained in Article 1(d). This described the last section of the Iraq–Najd boundary as running from Mukur 'to the Jebel Anazan (Anaza) situated in the neighbourhood of the intersection of latitude 32 degrees north longitude 39 degrees east where the Iraq–Najd boundary terminated' (Report of the Iraq Administration 1922–3).

Further progress on boundary making was hampered by the deterioration of relations between Ibn Saud and Husain bin Ali, the Hashemite *Sharif* of Mecca. A series of meetings in Kuwait designed to settle remaining territorial issues in northern Arabia (the abortive Kuwait Conference) in late 1923 and early 1924 failed to achieve anything, and 1924 and 1925 were consumed by open warfare in Arabia, the consequences of which were Ibn Saud's occupation of the Hejaz, Husain bin Ali's flight into exile, and Britain's formal annexation of the southern area of Transjordan from Ma'an to Aqaba (Holden and Johns 1981: 83–6, Leatherdale 1983: 46–50).

It was not until late in 1925, with Ibn Saud's fortunes in the ascendancy, that Britain sought a new opportunity to reopen negotiations with him concerning the boundaries of the Najd with those of Transjordan and Iraq. Sir Gilbert Clayton, a special envoy with knowledge of the area and its tribes, was sent to negotiate with Ibn Saud, and the outcome was the Hadda Agreement of 2 November 1925.

Clayton had been briefed by Roland Vernon of the Colonial Office on British policy on the border. Britain was still in occupation of Kaf in the Wadi Sirhan, and considered Jebel Anaza with its tribal links to Iraq to be vital to the security of the Imperial route to the East. Clayton was told that if necessary he could yield Kaf to Ibn Saud, but not Jebel Anaza. As Vernon wrote:

such a frontier could involve the intersection of Najd territory between Iraq and Transjordan, and could place ibn Saud astride the imperial route to the east. *This cannot be permitted,* and in no circumstances should you assent to any extension of Najd territory to the north which could have the effect of separating Iraq from Transjordan.

(Vernon, quoted in Collins 1969: 79; Vernon's emphasis)

As with the Uqair Protocols, where Britain did not press its claims to al-Jauf in order to get an agreement on the Iraq–Najd boundary, so at Hadda; Kaf and most of the Wadi Sirhan was ceded by Britain in order to retain control of Jebel Anaza. As a result, and notwithstanding the authenticity of Ibn Saud's claims to Kaf,[6] he did not get a contiguous boundary with Syria.

The wording of the Hadda Agreement concerning the location of Jebel Anaza differs from Article 1(d) of the Uqair Protocols by referring only to the coordinates longitude 39 degrees East and latitude 32 degrees North, as the point where the borders of Transjordan, Iraq and the Najd meet, whereas the Uqair Protocols associated these coordinates with the location of Jebel Anaza. This difference in wording turned out to be crucial to a final settlement of the boundary line, because in 1927 it was discovered that Jebel Anaza was some 23 miles to the north-east of the coordinates referred to and would have given Ibn Saud more territory. This Britain and Iraq were not prepared to do.

THE BOUNDARY AGREEMENT OF 1932

In September 1927 the Council of Ministers in Iraq had agreed to move the process forward of fixing the Transjordan–Iraq boundary by defining it on the basis of natural features. The discrepancy between the coordinates designating the triangulation point between Transjordan, Iraq and the Najd, and the actual location of Jebel Anaza, was not discovered until this time ('The Boundary between Iraq and Transjordan', by C.J. Empson, adviser to the Ministry of the Interior, Iraq, 4 November 1940, FO 816/20).

It had been decided to mount a proper boundary survey to fix the border and the ensuing 'compass traverse' (by whom it is not clear), and a reconnaissance by Major A.L. Holt for the Iraq government discovered the discrepancy between where Jebel Anaza was and where it was thought to be (Hubert Young, Acting High Commissioner, Iraq, to Colonial Secretary, 27 January 1932, CO 732 56/6).

This discrepancy produced considerable anxiety in the Colonial Office regarding the safety of the oil pipeline under construction from Kirkuk to Haifa across Transjordan, and reinforced the view that a formal boundary survey was needed. However, such a survey did not take place because of a war that broke out between Ibn Saud and the Ikhwan leader, Faisal al-Duwish in the vicinity of the boundary zone, and which lasted until 1930 (Troeller 1976: 238–9; Habib 1978: 6–8).

This delay in settling the Transjordan boundary threatened to prevent Iraq's admission into the League of Nations, due in October 1932. Hubert Young wrote to the Colonial Secretary to point out that if Iraq did not have defined boundaries, the League could reject, or at least delay, Iraq's application (Young to Colonial Secretary, 27 January

1932, CO 732 56/6). Here Britain also took steps to reaffirm the Iraq–Kuwait boundary (see Chapters 4 and 5).

Young took the view that a proper boundary survey would not only take time, but would inevitably lead to politically difficult correspondence being opened up with Ibn Saud. Ibn Saud was known to be unhappy with the Hadda Agreement and with the loss of Jebel Anaza. In drawing the Colonial Secretary's attention to the discrepancy between the Uqair Protocols which referred to Jebel Anaza, and the Hadda Agreement which did not, Young observed, 'it is necessary to determine the exact position of Jebel Anaza and to remove the inconsistency between the Uqair Protocols and the Hadda Agreement' (Young to Colonial Secretary, 27 January 1932, CO 732 56/6).

To overcome the time problem and the possible delay to Iraq's admission to the League, Young proposed a new definition of the location of the triangulation point based on the Uqair Protocols and the Hadda Agreement which was consistent with what was already known of the geography of the boundary zone. General terms were to be used until a formal boundary survey had been conducted.

The Colonial Secretary, Philip Cunliffe-Lister, advised the High Commissioner in Iraq that the Uqair Protocols took precedence over the Hadda Agreement, but that any new description should not draw attention to the discrepancy between the two documents. He also felt that the new boundary could be agreed without reference to the League and that the final delimitation could take place after Iraq was independent. A way of avoiding difficulties with the League in advance of a boundary agreement would be to adopt the wording of the Franco-British Convention of 1920. This would suggest that the boundary had already been defined (Colonial Secretary to High Commissioner, Iraq, 26 April 1932, CO 732 56/6).

The argument that the boundary had, in principle, already been defined in 1920, meant that only a formal agreement rather than a detailed boundary demarcation between Transjordan and Iraq would be required at this time. So the Colonial Secretary suggested a procedure whereby the Iraqi Prime Minister would write to his counterpart in Transjordan proposing that their boundary be agreed. The Chief Minister in Amman would then write to the British Resident in Transjordan informing him of Iraq's request and seeking the British government's approval. The British Resident would reply giving his approval on behalf of the British government, and the Chief Minister would then write to the Prime Minister relaying this approval.

The suggested procedure was strictly adhered to, and the description of the boundary, thrashed out in correspondence between officials in London, Baghdad and Amman, was eventually presented in a letter from Iraq's Prime Minister, Nuri al-Said, to the Chief Minister of Transjordan, Abdullah al-Sarraj, dated 31 July 1932. The key passage read:

> The frontier between Iraq and Transjordan starts in the south at the point of junction of the Iraq–Najd frontier and the Transjordan–Najd frontier and ends in the north at that point on the Iraqi–Syrian frontier and the Transjordan–Syrian frontier, as ultimately delimited which is nearest to the summit of Jebel Tenf. Between these two points the frontier follows where possible prominent physical features, provided that it shall not diverge more than five kilometres from a straight line between these two terminal points.[7]

The rest of the procedure then followed Cunliffe-Lister's recommendations. Abdullah al-Sarraj wrote to British Resident Sir Henry Cox on 9 August to seek the latter's approval of Iraq's request. Cox replied in the affirmative on 13 August; al-Sarraj then gave the final confirmation to Nuri al-Said in a letter dated 16 August 1932. Through this exchange of letters the boundary was agreed.

EVENTS SUBSEQUENT TO 1932

The final text of the Transjordan–Iraq boundary agreement deliberately avoided mentioning Jebel Anaza, partly so as not to draw attention to the discrepancy between the Uqair Protocols and the Hadda Agreement, and partly to meet the requirement that the wording should be a modified version of the Franco-British Convention of 1920. Britain did not want it to appear to the League as if the boundary had been redefined. The cause of this confusion over the location of Jebel Anaza was the War Office's 1918 map of Asia (British Embassy to Ministry of Foreign Affairs, Baghdad, 29 December 1939, CO 831 55/2) (see Figure 8.3).

A proposal by the War Office to produce a new map replacing that of 1918 caused concern to the Counsellor in the Foreign Office, Sir George Rendel. The Franco-British Convention of 1920 had been based on the old map, and any changes which exposed the errors of geography would cause a major embarrassment. Thus, he advised that a new map could only be published if it showed the contours of the land as featured in the 1918 map. He added:

> If a new sheet is published by the War Office with the frontier

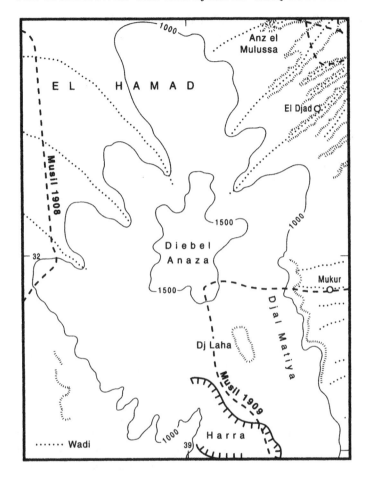

Figure 8.3 The War Office map of Asia, 1918 (detail)

shown in accordance with the coordinates mentioned in the Hadda agreement, our case will be very much weakened, and the fact that the War Office showed the frontier in this way on the new map will always be used against us. It is really essential that this should be prevented.

(Rendel's Minute of 9 September 1937, CO 732 76/2)

A border survey team was assembled in 1940. Comprising representatives of Transjordan, Iraq and Saudi Arabia, the team was instructed to do two things: (a) fix the exact position on the ground of the intersection of parallel 32 degrees North and meridian 39 degrees East; and (b) fix the coordinates of the peak of Jebel Anaza (Sir Basil Newton [Ambassador to Iraq] to Lord Halifax [Foreign Secretary], 2 February 1940, CO 831 55/2).

The border team discovered that Jebel Anaza consisted of two peaks, eight miles apart from each other. The tallest peak was 940 metres in height, the lower one 935 metres. The Saudi representative wanted the whole mountain to mark the boundary, but the teams representing Transjordan and Iraq preferred the tallest of the peaks because it was closest to the coordinated 32 degrees North and 39 degrees East. On this point the Saudi representative dissented ('The Frontiers between Transjordan and Najd and Transjordan and the Hejaz, Eastern Department, Foreign Office, September 1939', quoted in Schofield and Blake 1988: 548).

CONCLUSION

The Transjordan–Iraq boundary was fixed by Britain as a consequence of its acquisition of the mandates of Iraq and Transjordan under the League of Nations. The League's requirement that member states have defined boundaries was met in Iraq's case in 1932 because in that year the northern terminal was demarcated by the Syria–Transjordan Boundary Commission, although the southern terminal was only worked out on paper.

Procedurally, then, the geographical process of boundary evolution which is said to pass from allocation through delimitation to demarcation was only partially met in the case of Transjordan and Iraq, the southern terminal reaching only the stage of delimitation (Drysdale and Blake 1985: 77). This is not to imply, however, that many other boundaries in the region were demarcated before the Second World War. This was palpably not the case. There were two obstacles to the completion of the boundary-making process: the basic locational error in the War Office map of 1918, and the need to contain Ibn Saud of Arabia so as to protect the Imperial route to the East.

British officials were aware that the texts of the existing treaties gave a false or at least ambiguous location of the southern triangulation point of the border when the 1932 agreement was reached to settle the boundary line between Transjordan and Iraq.

At the heart of this process lies the problem of boundary formation for the modern state at a time when such a concept was meaningless to the nomadic inhabitants of Transjordan, Iraq and the Najd.

The territorial boundaries that were agreed to in the 1920s and the 1930s did not serve the same interests for Britain as they did for Ibn Saud. Britain's actions were shaped by regional and international interests: the control of the Imperial route upon the East, and the economic exploitation of Iraq. Territorial boundaries were fixed to meet these interests and the effect they had upon the nomadic tribes of the areas concerned was of secondary importance.

For Ibn Saud, however, the long-term objective to integrate the Arabian tribes into a Wahhabi polity under his command was as important as the territorial dominance that implied. That is, not just control of the tribes, but also of their resources in what at that time was a poor land. In addition, the boundaries of Wahhabi authority were conceived in terms of an Islamic project, something Britain could never aspire to.

It was only after the integration of the Najd and the Hejaz into the Kingdom of Saudi Arabia that a process of political centralisation began to take place (Kostiner 1990: 233). In this respect, the period which followed the First World War in the Middle East can be recognised as a crucial moment in the region's history. What had been the provinces of an empire were, at the territorial level, reconstructed as separate states, with different rulers and political systems, and boundaries validated by a process of international law based in Europe. These externally imposed boundaries exploited existing divisions among the Arabs, even as they laid the basis for new ones.

NOTES

1 This is an expanded version of the paper presented to the Conference. I am grateful to Brian Birch and Malcolm Wagstaff for their comments on an earlier draft of this paper, and to the Geography Department at Southampton University, Carl Grundy-Warr and IBRU for giving me the opportunity to attend the Conference. I am also grateful to the Cartographic Unit, Southampton University, who prepared the maps for publication.

2 It should be noted that the boundary under discussion here was modified in its southern reaches by a Jordan–Iraq agreement of 1984. Both states seem satisfied with the revision of the border introduced by this agreement.

3 Since independence in 1946, Jordan has realigned its boundaries with Saudi Arabia and Iraq. In 1965, Jordan and Saudi Arabia agreed to realign the boundary, extending Jordan's coastline on the Gulf of Aqaba in exchange for territory inland. This retained the status of Jebel Anaza as the triangulation point of the Jordan–Iraq–Saudi Arabia boundary, but from this point the boundary ran in the first instance to the junction of the coordinates 32 degrees North 39 degrees East, before running south-west, whereas before

the boundary from Jebel Anaza ran in a straight line as far as Mudawarra. The text of the 1984 Jordan–Iraq boundary modification has not yet been made publicly available, though the delimitation introduced is depicted clearly in the US State Department's Office of the Geographer's Geographic Notes: number 13, March 1991: 2–4.

4 The settlement of the Transjordan–Syria boundary was considered earlier but was delayed, first by the Druze Rebellion in Syria (1925–7), and then by the problematic negotiations between Britain and French officials in London, Paris and Beirut that were not finally resolved until 1932. On the demarcation of the Syria–Transjordan boundary, see the Report of the British Resident to the High Commissioner, 14 August 1932, CO 732 54/2. The location of Jebel Khurgi and the modification of the wording of the 1920 Convention had already been remarked upon by Kenneth Blaxter in the Colonial Office on 18 May 1932, CO 732 56/6.

5 Note by Lawrence, 'Transjordan-Extension of Territory', 5 January 1922, CO 733 33. Winston Churchill as Colonial Secretary relied on Lawrence's arguments when communicating with the High Commissioner in Jerusalem, see his despatch to Sir Herbert Samuel of 8 February 1922, CO 733 33.

6 Kaf was considered important to Transjordan because there was a motor road running through it which connected Ma'an, Jafr and Bayir with Azraq; and because the wells and grazing grounds in the area were used by predominantly Transjordan-based tribes, the Bani Sakhr and the Rwala. However, Kaf and the surrounding villages were economically dependent upon the salt trade, and Ibn Saud claimed that Transjordan's border would sever villages from Kaf and disrupt the local economy. He also claimed Kaf as a 'right of war' after defeating Husain bin Ali (Collins 1969: 80–1).

7 For the complete set of letters exchanged between Baghdad and Amman constituting the boundary agreement, see the 'Report of His Britannic Majesty's Government ...' etc., for 1932.

REFERENCES

Abidi, A.H.H. (1965) *Jordan: A Political Study 1948–1957*, London: ASRA Publishing House.

Al-Azmeh, A. (1986) 'Wahhabite polity', in I. Netton (ed.), *Arabia and the Gulf: From Traditional Society to Modern States*, London: Croom Helm.

Baylson, J. (1987) *Territorial Allocation by Imperial Rivalry: The Human Legacy in the Near East*, University of Chicago: Department of Geography, Research Paper No. 221.

Bureau of Intelligence and Research, USA (1969) *Jordan–Syria Boundary*, International Boundary Studies No. 94, Washington, DC: Department of State.

—— (1970) *Iraq–Jordan Boundary*, International Boundary Studies No. 98, Washington, DC: Department of State.

Collins, R.O. (ed.) (1969) *Sir Gilbert Clayton: An Arabian Diary*, Berkeley and Los Angeles: University of California Press.

Drysdale, A. and Blake, G.H. (1985) *The Middle East and North Africa: A Political Geography*, New York and Oxford: Oxford University Press.

Habib, J.S. (1978) *Ibn Saud's Warriors of Islam*, Leiden: E.J. Brill.

Holden, D. and Johns, R. (1981) *The House of Saud*, London: Sidgwick & Jackson.

Khoury, P.S. and Kostiner, J. (eds) (1990) *Tribes and State Formation in the Middle East*, Berkeley, Calif.: University of California Press.

Klieman, A.S. (1970) *The Foundations of British Policy in the Arab World: The Cairo Conference of 1921*, Baltimore and London: The Johns Hopkins Press.

Kostiner, J. (1990) 'Transforming dualities: tribe and state formation in Saudi Arabia', in P.S. Khoury and J. Kostiner (eds), *Tribes and State Formation in the Middle East*, Berkeley, Calif.: University of California Press.

Leatherdale, C. (1983) *Britain and Saudi Arabia 1925–1939: The Imperial Oasis*, London: Frank Cass.

Luke, H.C. and Keith-Roach, E. (1934) *The Handbook of Palestine & Transjordan* (2nd edn), London: Macmillan.

Moore, J.N. (ed.) (1974) *Arab–Israeli Conflict, Vol. 3: Documents*, Princeton N.J.: Princeton University Press.

Netton, I. (ed.) (1986) *Arabia and the Gulf: From Traditional Society to Modern States*, London: Croom Helm.

Schechtmann, J.B. (1937) *Transjordan im Bereiche des Palastinaman-dates*, Wien: Dr Heinrich Glanz Verlag.

Schofield, R. and Blake, G. (eds) (1988) *Arabian Boundaries, Primary Documents: 1853–1957. Vol. 6: Saudi Arabia–Transjordan II*, Farnham Common: Archive Editions.

Shwadran, B. (1959) *Jordan: A State of Tension*, New York: Council for Middle Eastern Affairs.

Towle, P.A. (1989) *Pilots & Rebels: The Use of Air Power in Unconventional Warfare, 1918–1980*, London: Brassey's.

Troeller, G. (1976) *The Birth of Saudi Arabia: Britain and the Rise of the House of Saud*, London: Frank Cass.

Yapp, M.E. (1987) *The Making of the Modern Near East 1792–1923*, London: Longman.

9

LEGAL ASPECTS OF THE IRAQI SOVEREIGNTY AND BOUNDARY DISPUTES WITH KUWAIT

Maurice H. Mendelson

INTRODUCTION

'Some aspects of the Iraq–Kuwait sovereignty dispute are historic, whilst others are current'. (Mendelson and Hulton 1991)

When, in 1990, Iraq purported to annex the whole territory of Kuwait, it was in pursuance of a 50-year-old claim to sovereignty over the whole country, including its adjacent islands. That claim has now been brought to an end by a combination of military and legal means: military action by the coalition acting under the authority of the Security Council, and legal action by the Council's refusal to recognise the annexation (notably in SC res. 662 of August 1990), Iraq being forced in the end formally to abandon its sovereignty claim. Iraq had in fact previously acknowledged Kuwait's sovereignty and a rather vaguely described boundary, notably in a 1963 agreement entitled 'Agreed Minutes Regarding the Restoration of Friendly Relations, Recognition and Related Matters' signed at Baghdad on 4 October 1963 (Treaty No. 7063, 485 *UN Treaty Series* 321–29), but its validity was subsequently denied by Saddam Hussein.

The 1963 agreement reaffirmed, without reiterating, an earlier agreement of 1932 which defined the frontier between the two countries. However, that agreement, which itself simply repeated an earlier one of 1923, did not specify the line with any degree of precision. This has, in the past, been a source of contention between the two states and, indeed, alleged boundary encroachment by Kuwait was one of the grounds involved by Saddam Hussein to justify his invasion.

Consequently, when, on 3 April 1991, the Security Council laid down its terms for a cease-fire with Iraq in resolution 687, it demanded that Iraq (and Kuwait) respect the inviolability of the boundary and called on the UN Secretary-General to lend his assistance in its demarcation. In his report a month later (UN doc. S/22558, 2 May 1991), the Secretary-General reported that resolution 687 had been accepted and that he intended setting up a boundary commission comprising one Iraqi, one Kuwaiti and three neutrals (United Nations Iraq–Kuwait Boundary Demarcation Commission (UNIKBDC)). The Security Council approved his report (UN doc. S/22593 of 13 May 1991), and the members of the demarcation commission (UNIKBDC) were soon thereafter approved (UN Press Release SG/A/462, IK/20, 22 May 1991). They were: Professor Mochtar Kusuma-Atmadja (a lawyer and former Foreign Minister of Indonesia), chairman (Professor Kusuma-Atmadja was replaced as chairman by Greek jurist Nicholas Valticos in early November 1992); Mr Ian Brook (a Swedish surveyor); Mr William Robertson (a New Zealand surveyor); Ambassador Riyadh al Qaysi (an Iraq lawyer and diplomat); and Dr Tariq A. Razouki (a Kuwaiti lawyer and diplomat). The commission is empowered to decide by majority, and its decisions are final. The commission commenced its work in the summer of 1991, announced its decision on the Iraq–Kuwait land boundary during April 1992 (a line demarcated by November 1992) and, at the time of this volume going to press, has still to announce a final decision on the water boundaries between the two states.

Aspects of the proceedings of the Security Council up to and including July 1991 could conceivably give rise to controversy in the future. There is no doubt as to the Council's authority, under the UN Charter, to take the action needed to maintain or restore international peace and security; but the Council's affirmation of the binding force of the 1963 agreement – whose validity had been questioned by Iraq – is, I think, unprecedented. In my view, the Security Council's power to make binding decisions with regard to the maintenance of international peace and security under Chapters I, II, VI and VII of the UN Charter constitute sufficient authority for this action: Iraq needed to be ordered to respect the international boundary, and for this purpose the Council had to determine what the boundary was. The Agreed Minutes, a treaty registered with the UN under Article 102 of the Charter and un-challenged for many years by Iraq, afforded a good basis for the decision. As I shall show below, Iraq's denial of the validity of that agreement was unconvincing; and the Council, which briefly considered its objections before adopting resolution 687, evidently agreed. It is, however, perhaps unfortunate that the validity of the 1963 agreement could not be determined in a more formal manner, for it seems that Iraq has not, even now, really abandoned its rejection of the Agreed Minutes. Its letter of 6 April 1991 to the Secretary-General accepting resolution 687 (UN doc. S/22456, annex) was so full of reservations about this and other matters that the latter was forced to get Iraq's ambassador to the UN to confirm his country's unconditional acceptance of the resolution (UN docs S/22480 and 22485 of 11 April 1991). In its letter of 6 April, Iraq asserted that resolution 687, by imposing a particular boundary, was an 'unprecedented assault' on the sovereign rights of a state, contrary to the Charter and international law and custom; that the 1963 agreement was not in force; and that Iraq was participating in the demarcation process only under duress. So far as concerns the demarcation itself, Iraq raised questions about the possible independence of the future 'neutral' members of the Commission, since it would have no hand in their selection, and complained about certain other matters.

One can see here the seeds of a possible future conflict: if Iraq regains anything like its former strength and outside forces are unable or unwilling to prevent it, Iraq may some day revive its former claim, asserting that settlements imposed by force are null and void under international law and that the demarcation process was unfair. I am not saying that such arguments would be well founded; just that an opening has been left.

For all of these reasons, the background to the dispute is therefore still relevant, not just of interest to historians. We shall therefore look briefly at the history of Kuwait and the legal validity of Iraq's claims to the whole territory of Kuwait (including its islands), before turning to the question of the precise demarcation of the boundary.

HISTORY OF THE DISPUTE

In the nineteenth century both countries were part of the Ottoman Empire. What is now called Iraq, namely, the *vilayets* or provinces of Mosul, Baghdad and Basra, certainly were firmly under Turkish sovereignty. Kuwait was regarded by the Ottomans as attached to the *vilayet* of Basra, though they had never occupied it or exercised firm jurisdiction over it. The al-Sabah ruler of Kuwait held the Turkish title of *qaimmaqam* (prefect or governor) of his country. In 1896 Sheikh Mubarak al Sabah murdered his reputedly pro-Turkish brothers and, since the Sublime Porte would not recognise his independence, he asked Great Britain to grant him protection. In 1899 a secret agreement was

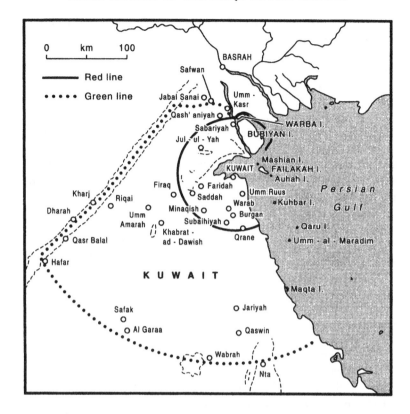

Figure 9.1 Red and Green lines introduced to define Kuwaiti territory after the
Anglo-Ottoman Convention, 29 July 1913

Source: Richard Schofield 1991, *Kuwait and Iraq: Historical Claims and Territorial
Disputes*, London: Middle East Programme, Royal Institute of International Affairs,
p. 47, by permission of the publishers.

Note: Transliterations are consistent with those appearing on original Foreign Office
Research Department map (to be found in PRO file: FO 371/114644) of which the above
is a copy.

entered into, whereby Britain offered him its protection, in return for
which he undertook to cede no territory and not to receive any foreign
representative without British consent. This did not, however, prevent
him from accepting from the Sublime Porte the title of *qaimmaqaam*
when the latter came round to recognising him as sheikh. Shortly afterwards, with the encouragement of the British, Sheikh Mubarak claimed
the islands of Warba and Bubiyan because of their strategic importance
(Schofield and Blake 1988: 35–52).

The Ottomans' influence in Arabia was on the wane, and in 1913 they entered into an agreement with the UK in which, amongst other things, they recognised the 1899 agreement and the autonomy of the al-Sabah within a 40-mile radius around Kuwait town, as well as a rather weaker authority in an area beyond this. The inner zone is marked 'Red Line' on Figure 9.1; the outer, 'Green Line'. The British, for their part, acknowledged formal Ottoman suzerainty there, and undertook not to establish a formal protectorate so long as the *status quo* continued. The outbreak of the First World War prevented the Convention from being ratified, and in 1914 the Ruler of Kuwait revolted against the Turks, throwing in his lot with the British, who promised recognition as an independent government under their protection (though not technically establishing a protectorate) (Schofield and Blake 1988: 246).

After the dissolution of the Ottoman Empire in 1918, it was agreed that the Mesopotamian provinces of Baghdad, Mosul and Basra should form the self-governing state of Iraq, and that Great Britain should administer it under a League of Nations mandate until it was ready for independence. Under the Treaty of Lausanne of 1923 Turkey renounced all of its former territory outside its present borders with the possible exception of Mosul province, not formally relinquished until the conclusion of an Anglo–Iraqi–Turkish treaty of 1926.

Great Britain decided that the boundary between Iraq and Kuwait should be settled, and this was done in 1923 by an exchange of letters involving the British Political Agent in Kuwait, the British High Commissioner in Baghdad and the Ruler of Kuwait (Schofield and Blake 1988: 219–29, with details of the boundary below). Warba and Bubiyan, as was the case in the Anglo-Ottoman Convention of 1913, were amongst the islands included on the Kuwaiti side.

Great Britain gave up the mandate in 1932, and on 3 October Iraq became an independent sovereign state and was admitted to the League of Nations. Meanwhile, at the instigation of the British, who acted as intermediary, an agreement was reached between the two countries to reaffirm the boundary by means of an exchange of letters in July to August from the Iraqi Prime Minister, Nuri al Said, and the Ruler of Kuwait (Schofield and Blake 1988: 372, 376).

On 19 June 1961 the British agreements with Kuwait were brought to an end, and the latter applied to join the United Nations as a fully independent state. Six days later the Iraqi leader, General Kassem, claimed sovereignty over the whole of Kuwait, claiming that Ottoman sovereignty had not been lost before Iraq had succeeded to Turkish rights over the province of Basra. Following reports that Iraqi troops were moving southwards, British and Saudi troops went to Kuwait's defence to be replaced shortly afterwards by an Arab League defence force. Following the overthrow of General Kassem in February 1963 and his replacement by President Aref, relations improved; and on 4 October an agreement was entered into which I have already mentioned – the 'Agreed Minutes' – by which Iraq, amongst other things, 'recognise[d] the independence and complete sovereignty of the state of Kuwait and its boundaries as specified in the letter of the Prime Minister of Iraq dated 21 July 1932 and which was accepted by the Ruler of Kuwait in his letter dated 10 August'. However, it became apparent that this acceptance did not extend to the precise delimitation of the boundary, and disagreement and discussions between the countries dragged on for years. With the development of the Iraqi port of Umm Qasr in the Khor Zubair as an alternative egress to the Gulf to the Shatt al-Arab, the islands' controlling position became more important to Iraq; and with the latter's development of the Rumaileh oilfield, the precise location of the frontier became a matter with great economic implications. With the outbreak of the Iraq–Iran war, in which Kuwait leaned towards the former, relations improved: but with the ending of that war, Iraq stepped up its complaints and claims, adding to them a complaint that Kuwait, by overproduction of oil, had deliberately undermined the Iraqi economy and that it had also stolen vast amounts of oil from the southern section of the Rumaileh oilfield.

On 2 August 1990, Iraq invaded and overran Kuwait. Initially President Saddam Hussein claimed that he had been invited in by internal Kuwaiti revolutionaries, and a 'Provisional Free Kuwaiti Government' was proclaimed, though it was widely denounced as a sham. On 8 August, the day after the PFKG proclaimed Kuwait a republic, Iraq formally annexed the whole country, though allegedly at the request of the new government. All foreign embassies in Kuwait were ordered to move to Baghdad but many refused to comply. The occupation was brought to an end, under Security Council auspices, by Operation 'Desert Storm' and by Iraq's acceptance of the invalidity of the occupation.

THE LAW

What are we to make of Iraq's claims? In the first place, it seems to me that there is a good case for saying that Kuwait had established its independence from the Ottoman Empire before the creation of Iraq, if not in 1899 then between 1913 and 1918. If so, then Iraq's claim to sovereignty over the whole country by means of state succession to Turkey's rights (which in any case may have been less than sovereignty) falls by the wayside, though not necessarily its claim to Warba and Bubiyan. Secondly, and even if that is wrong, recognition by the competing claimant can cure any defects in title. And, as we have seen, there was recognition of Kuwait's separate existence and of its title over the islands in 1923, 1932 and 1963. Baghdad objects that in 1923 and 1932 Iraq was not independent, and so recognition by the British High Commissioner in 1923, and by Nuri al Said in 1932, does not count. But is this correct? State practice, arbitral awards and decisions of the International Court of Justice establish that internal administrative boundaries of a single colonial power, boundaries agreed between colonial powers, and boundaries between colonial powers and independent states, are to be respected after independence. The 1923 agreement seems to fall within the first or third categories and, on the Iraqi showing, so, essentially, does the 1932 agreement. The fact that Iraq was not independent in 1923 or 1932 does not, therefore, seem significantly to affect the issue. Kuwait's independence has also, of course, been recognised by many other states and international organisations, including the UN and the Arab League.

Iraq also claims, however, that the 1932 and 1963 agreements were invalid because the approval of the National Council and National Revolutionary Council, respectively, was not obtained. Although I am not an expert in Iraqi constitutional law, my research suggests that, as a matter of domestic constitutional law, this analysis is not correct; but even if it were, there are serious problems in international law. Article 46 of the Vienna Convention on the Law of Treaties, 1969 (which is not retrospective but in this respect seems to be declaratory of the position in customary law) states as follows:

1   A state may not invoke the fact that its consent to be bound by a treaty has been expressed in violation of its internal law regarding competence to conclude treaties as invalidating its consent unless that violation was manifest and concerned a rule of internal law of fundamental importance.

2   A violation is manifest if it would be objectively evident to any state conducting itself in the matter in accordance with normal practice and good faith.

Whether the proviso has been satisfied in this case is a question of fact; but the burden of proof is on the state alleging invalidity, and it is not an easy burden to discharge, as the arbitral award of 31 July 1989 in the case of *Guinea-Bissau/Senegal* shows. Furthermore, after so many years of silence, Iraq could well be stopped from now challenging the validity of the 1963 agreement. And, finally, since Iraq's acceptance of Security Council resolution 687 had the express endorsement of the National Revolutionary Council (UN doc. S/22480, 11 April 1991), it could be argued that any deficiency in that regard has now been cured.

So much for Iraq's claims to sovereignty over the whole of Kuwait and its islands, and to its challenges to the legality of the instruments specifying the frontier. I now turn to the main question which has faced the boundary demarcation commission: What exactly is the course of the boundary?

THE BOUNDARY

The relevant part of the boundary was specified for the first time in the unratified Anglo-Turkish Convention of 1913. The relevant provision states (in my translation from the original French):

> The line of demarcation leaves the coast at the mouth of the Khor Zubair in a northwesterly direction and passes immediately south of Umm Qasr, Safwan and Jebel Sanam in such a way as to leave these places and their wells to the *vilayet* of Basra; when it reaches the Batin it follows it towards the south-west as far as Hafr-el-Batin, which it leaves to Kuwait.

The 1923 exchange of letters between the Sheikh of Kuwait, the British Political Agent in Kuwait and the British High Commissioner in Baghdad was based on this line, but was not in precisely the same terms. It also reverses the order of the description, defining the frontiers as running:

> From the intersection of the Wadi el-Audja with the Batin and then northwards along the Batin to a point just south of the latitude of Safwan; thence eastwards passing south of Safwan well, Jebel Sanam and Umm Qasr, leaving them to Iraq and so on to the junction of the Khor Zubair with the Khor Abdullah.

No map was annexed to the agreement. The 1932 exchange of letters between Iraq's Prime Minister and the Sheikh of Kuwait 'reaffirmed' this frontier and is in virtually identical terms; again, no map was annexed. The 1963 Agreed Minutes, by which Iraq seemingly abandoned its pretensions to sovereignty over Kuwait and reaffirmed the boundary, simply referred back to the 1932 exchange of letters.

What, then, are the questions with which the commission has had to deal? They appear to be:

1   The meaning of the term 'along the Batin'.
2   The location of the point 'just south of the latitude Safwan'.
3   The course of the boundary to the junction of the Khor Zubair and the Khor Abdullah.

There is also a need to determine the course of the boundary within the Khor Abdullah. This is not mentioned in any of the treaties, but the commission may possibly consider that it has the authority to settle it, at least as far as the baselines of the territorial sea. There is also a need for the territorial sea and continental shelf boundary of the two states to be delimited, but it seems less likely that the commission will undertake that task, which may well be outside its remit.

There have been attempts in the past to add greater precision to the definition of the boundary. The British on behalf of Kuwait, and later Kuwait itself, pressed for this. Thus in 1940 the British government communicated their interpretation of the the meaning of the key terms in a letter from the British Ambassador in Baghdad to the Iraqi Foreign Minister; and in a *note verbale* of 18 December 1951 to the Foreign Minister this interpretation (with one difference which I shall revert to) was reiterated, together with a request for Iraq's views (Foreign Office 1987: 246). Iraq, however, refused to consider the question of the boundary until it had obtained the transfer by cession of sovereignty or by lease of Warba and/or Bubiyan, which were of strategic importance to it, inasmuch as they controlled the only egress to the Gulf apart from the much disputed Shatt al-Arab. For the British and the Kuwaitis, this was the wrong order of proceeding.

Before going into the details of these questions about the course of the boundary, I should perhaps briefly mention the question of the map which Security Council resolution 687 specifically singles out as part of the 'appropriate material' to be taken into account in the demarcation. This is in fact a series of maps on a scale 1:50,000 which were made by the British Ministry of Defence in 1988 and revised in 1990, and which had been transmitted to the Secretary-General by the British government in a letter dated 28 March 1991 (UN doc. S/22412). This has caused a great storm in a teacup, with the Iraqis protesting that they had no hand in its production and did not recognise it. However, it is to be noted that the resolution only specifies the maps as *one* of the items to be considered: it does not make them in any way dispositive. And in fact, the letter accompanying these maps drew attention to the fact that they bear on their face the note: 'Maps produced under the direction of the Director-General of Military Survey are not to be taken as necessarily representing the view of the UK Government on boundaries or political status.' It rather looks as if the Security Council has attached more emphasis than was intended to maps forwarded by the British government simply in order to be helpful – the maps being both recent and large scale.

Let us look a little more closely, then, at the issues the boundary commission have had to confront – though I have necessarily had to condense or miss out a large part of a complicated legal and factual analysis, and will concentrate on raising the issues rather than suggesting the answers, most of which depend on a mass of factual information which was not yet publicly or readily accessible during the summer of 1991.

'Along the Batin'

What exactly is meant by the phrase 'along the Batin'? The Batin is a long and fairly wide *wadi* or valley, which is dry except when it rains and is used for grazing and recreation. The questions which arise in this regard are whether the boundary here should run in a straight line or whether, in view of the discovery after 1923 that the course of the Batin is rather sinuous, the line should follow that course; and, if so, whether it should run along one of the banks, the *medium filum aquae* (median line) or the *thalweg* (deepest navigable channel).

The unratified Anglo-Turkish Convention is the historic origin of the boundary, and it says that the line follows the Batin, which rules out a straight line even if one was drawn on the small-scale map annexed. But for technical reasons it is doubtful that it can be treated as part of the drafting history of the later agreements so as to constitute a 'subsidiary means' for their interpretation, as laid down in the Vienna Convention on the Law of Treaties. Nor does there seem to be any contemporary or subsequent document or agreement by both the parties establishing their agreement as to the meaning of the phrase. One is probably therefore thrown back onto principles of customary international law. The British position, communicated to Iraq in 1937, 1940 and 1951 is that the line must follow the *thalweg*. This is the usual line of demarcation of navigable rivers; but for *non*-navigable rivers the normal principle is equal division, and the Batin is not – or at any rate is not usually – navigable, being dry for most of the year. At the moment, the only justification I can see for using the *thalweg* would be the difficulty of defining what is the Batin, it being a relatively shallow depression surrounded, I believe, by unstable banks and shifting sand. Interestingly, the line on the British Military Survey maps to which Security Council resolution 687 refers does not seem to follow the deepest channel at all points, so far as I could tell from a brief inspection, and bearing in mind that I am not a geographer. I also understand that in some places the bed of the Batin is flat and gravelly, so that there is no deepest channel, and that in such cases the midway point between the highest points on the banks was used. Where there is no *thalweg* it certainly seems sensible to use the middle of the watercourse. But it also looks as if, at some points on the Batin, the Military Survey line crosses small hill-like features, which is not really consistent with the idea of following the deepest channel.

In any case, the fundamental question remains: is the British proposal of a line following the *thalweg* (where there is one) necessary under the treaties or appropriate?

According to a United Nations press release (IK/34 of 16 July 1991), the commission approved the 'methodology to be applied in demarcating the western portion' of the boundary, but I had, as of July 1991, no further information as to what this means in the context of the present discussion.

'Just south of the latitude of Safwan'

Secondly, what is meant by the expression in the 1923 and 1932 agreements: 'just south' of the latitude of Safwan, which marks the point at which the frontier should leave the Batin and turn eastwards? Similarly, what is meant by 'south of Safwan well', which is the point through which the line passes on its course to the east? This, of course, raises a question of fact as to what the parties had in mind. There is some suggestion that the frontier here was originally marked by a large notice-board located one mile south of Safwan, which had been erected by the Ruler of Kuwait and the British Political Agent in Kuwait in about 1923. This was apparently removed, however, by the Iraqis in 1932, restored by them, but once again removed in 1939 by persons unknown. In June 1940, the Political Agent had it replaced in the presence of an Iraqi frontier official, giving rise to a protest by the Iraqis that the new board had been erected at a point far from the site of the old one – at a distance of 250 metres within Iraqi territory. They removed it and it was not subsequently replaced, leading to suggestions that it should be re-instated or that some other point, selected by reference to coordinates or to a fixed landmark, should be used instead. At one stage, the most southerly palm tree in Safwan was considered for the landmark, but matters were complicated by the Iraqis putting in new plantations south of Safwan during or immediately after the Second World War.

In their proposal to Iraq in 1940, the British authorities suggested the point at which the notice-board had stood until 1939, but did not specify its location. The 1951 proposal was more precise but possibly specifies a different point – namely, 1,000 metres due south of the building which, on 25 June 1940, was used as the customs post at Safwan.

This issue gives rise to numerous and complex questions of fact and law, such as: how far can the Anglo-Turkish Convention, which uses the phrase 'immediately south', be used to interpret the phrase 'just south'?; were the two sides ever in agreement that the old notice-board (or any other point) *did* mark the frontier?; and, where was that board actually located? A particularly interesting issue, but one which I have no time to develop here, is whether, having put forward one interpretation in 1940, the British authorities were estopped (i.e. precluded in law) from putting forward another in 1951. And were they, or their successors (so to speak), the Kuwaitis, precluded from putting forward a different interpretation after 1951? (For instance, there is a suggestion that the point 1,000 metres south of the Safwan customs post, which was specified in 1951, is well to the north of where the line runs on the British Military survey. The latter appears to pass approximately 1.6 km south of Safwan.) The question of estoppel is potentially very important in all sectors of this boundary.

Already by mid-July 1991 it had been announced that UNIKBDC had agreed on the 'approximate location of the boundary in the area of Safwan in the northern section' (UN press release IK/34 of 16 July 1991).

The line east of Safwan to the Khors

A third issue concerns the course to be followed by the boundary in the segment from Safwan to the Khors. The language of the 1923 and 1932 agreements is that after the point 'just south of the latitude of Safwan' the line runs 'eastwards passing south of Safwan well, Jebel Sanam and Umm Qasr, leaving them to Iraq and so on to the junction of the Khor Zubair with the Khor Abdullah'. Does the boundary run in a straight line from the point south of Safwan to the junction of the Khor Zubair and Khor Abdullah, or in a direct line to a point immediately south of

153

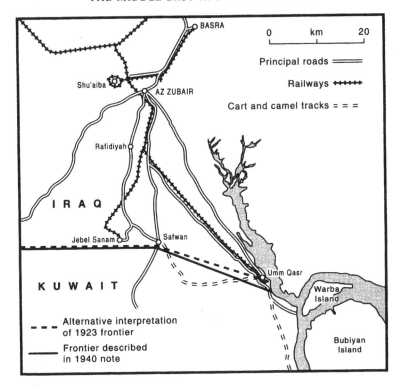

Figure 9.2 Disagreement between the Foreign Office and the Government of India during 1941 and 1942 on where the land boundary should terminate on the Khor Zubair

Source: Richard Schofield 1991, *Kuwait and Iraq: Historical Claims and Territorial Disputes*, London: Middle East Programme, Royal Institute of International Affairs, p. 88, by permission of the publishers.

Note: Transliterations are consistent with those appearing on original Foreign Office Research Department map (to be found in PRO file: FO 371/114644) of which the above is a copy.

Umm Qasr, then turning south-east to the junction of the Khors? (See Figure 9.2.)

Originally, and due it seems to an error of geography, the British authorities thought that a straight line to the Khors would run about 1 mile south of each of the named locations until it met the Khors, and it was therefore a straight line which they proposed to Iraq in 1940. They later realised their mistake, and the India Office argued strongly that the line should not turn to the south until after Umm Qasr. The difference

154

concerned a small wedge of territory, but potentially an important one as regards oil and the development of the port of Umm Qasr. However, the Foreign Office decided that it could not now go back on its 1940 proposal, and the line put forward in 1951 was the same one (Schofield 1991: 93–5).

Amongst the questions which spring to mind here are: firstly, does or does not the 1932 definition require a single straight line?; secondly, if it is ambiguous, can the words of the Anglo-Turkish Convention, which support an affirmative answer, be used as an aid to interpretation?; and thirdly, is Kuwait estopped from putting forward a line different from that proposed by Great Britain in 1940 and 1951?

The line on the British Military Survey map sent to the Security Council is straight: it does not follow the course proposed by the India Office. (Whereas the British map shows the boundary as running through Umm Qasr airport, a captured Iraqi military map which I have seen puts the line well to the south of it.)

MARITIME DELIMITATION

As I mentioned, two further possible issues concern the delimitation in the Khor Abdullah, and the maritime boundary (of the territorial sea and the continental shelf) between the two countries in the Gulf. However, these matters are not dealt with at all in the treaties; they are possibly outside the terms of reference of the boundary demarcation commission; and the commission had not by mid-summer 1991 yet made up its mind whether to consider them. I understood, though, that it may delimit up to the baselines from which the territorial sea (and continental shelf) are measured. Presumably, it will follow the *thalweg*. I do not propose to go into the questions of maritime delimitation here, though they are certainly interesting and ought to be resolved sooner rather than later in order to remove further causes of friction between the two countries.

The body which has been established with the help of the UN Secretary-General is a boundary commission, not an arbitral tribunal whose main function is to make a legal determination. It has a practical job of demarcation to do (that is, marking the line on the ground); some of its members are non-lawyers; and it may quite possibly get impatient about some of the legal subtleties I have outlined (not to mention others that I have not had the time to allude to here). But in my submission, it is important for both parties and the commission to deal properly with the legal aspects of the case – and I do not say that merely because I am an international lawyer who would be disappointed if these fascinating questions were not ventilated and decided. To my mind, there is a much more important reason why the legal issues should be properly confronted.

We had already seen by late spring 1991 that Iraq had begun to sow the seeds for a possible challenge to the decisions of the UNIKBDC. For the sake of stability in the region it is vitally important that the boundary the commission lays down can stick: we have seen too many instances of boundaries which have apparently been settled by a tribunal being subsequently challenged on legal grounds (whether spurious or not). It is therefore, to my mind, vital that the determination by the commission be as convincing as possible, not only technically and factually, but legally too.

Let us hope that this will be the case.

POSTSCRIPT

On 16 April 1992, after nearly ten months of deliberation, UNIKBDC made public its decision on the Iraq–Kuwait land boundary (IK/101/16, 16 April 1992). This made the following provisions for the disputed/unclear sections of the boundary enumerated and outlined above:

1   The boundary along the Batin followed the *thalweg* of that feature.
2   The 'boundary south of Safwan shall be located at a distance of 1430 metres from the south-west extremity of the compound wall of the old customs post along the old road from Safwan to Kuwait'.
3   The section from the point south of Safwan ran in a straight line to the tri-unction of the Khor Zubair, Khor Shetana and Khor Sabiya but deviated in such a way so as to leave the whole of the former water inlet to Iraq.

It will be noted therefore that in locations 1 and 3 the UN decision effectively followed Britain's 1951 demarcation proposal. The land boundary was demarcated by permanent pillar by the end of November 1992.

REFERENCES

Foreign Office (1953) *Historical Summary of the Events in the Persian Gulf Shaikhdoms and the Sultanate of Muscat and Oman, 1928–1953*, ('PG53'), Farnham Common: Archive Editions (reprint).

Mendelson, M.H. and Hulton, S.C. (1991) 'La revendication par l'Irak de la souveraineté sur le Koweït', *Annuaire français de droit international*, 36, 923–56.

Schofield, R. (1991) *Kuwait and Iraq: Historical Claims and Territorial Disputes*, London: Royal Institute of International Affairs.

—— and Blake, G.H. (eds) (1988) *Arabian Boundaries: Primary Documents 1853–1957*, Farnham Common: Archive Editions (volumes 7–8).

10

THE HISTORICAL PROBLEM OF IRAQI ACCESS TO THE PERSIAN GULF

The interrelationships of territorial disputes with Iran and Kuwait, 1938–90

Richard N. Schofield

INTRODUCTION

As the focus of this short discussion is the effect of the Shatt al-Arab dispute and Iranian–Iraqi relations more generally on the prosecution of Iraqi demands for control over the Kuwaiti islands of Warba and Bubiyan, a few words of introduction must be said about the basic nature of Iraqi claims to these features. Increasingly over the last half-century, Iraq has perceived itself as squeezed out of the Gulf (Figure 10.1). In an attempt to redress this situation, Iraq has endeavoured, with remarkable consistency since 1938, to modify the existing Kuwait–Iraq boundary as defined by diplomatic exchanges of 1923 and 1932 and confirmed by the Kuwait–Iraq agreement of October 1963, so as to improve its limited access to the waters of the Gulf. This has typically involved requests for the cession or lease of the strategically important Kuwaiti islands of Warba and Bubiyan, a move which would give the Baghdad government undivided control over the approaches to the Khor Zubair, on which Umm Qasr port is situated. Despite signing the 1963 'Agreed Minutes ...', apparently recognising the boundary delimitation on its own merits, Iraq has consistently demanded satisfaction on the islands question before any consideration of demarcating the land boundary would be entertained. Conversely, Kuwait, which has tenaciously resisted all suggestions that it might cede or trade portions of its northern land and islands territories, has traditionally refused to consider leasing Warba and Bubiyan unless Iraq first agreed to the

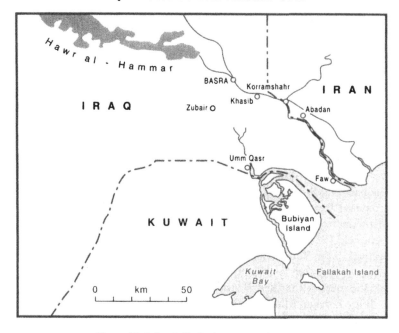

Figure 10.1 Iraq's limited access to the sea

demarcation of the existing land boundary. A solution to the border and islands question had long proved elusive and entrenched in this familiar deadlock before Iraq's disastrous action of invading Kuwait on 2 August 1990.

Iraq's awareness of its geographically disadvantageous position on the Gulf and its resultant frustration at not being able to play a substantive role there is clear from various formal and informal statements made over the years. Iraq has been described as 'a big garage with a very small door' (al-Mayyal 1986) or Kuwait 'the cork in the bottle' that is Iraq, while Iraqi government officials have, on more than one occasion, stated that Iraq could not possibly be a Gulf state unless it owned Warba and Bubiyan.[1]

INITIAL IRAQI DEMANDS FOR THE CESSION OF WARBA AND BUBIYAN, 1938–40

In March 1938 the Iraqi Foreign Ministry intimated that they would like to have an alternative outlet to Gulf waters other than the Shatt al-Arab, preferably an Iraqi-controlled port on Kuwait Bay.[2] The various departments of the British government met frequently thereafter to discuss the merits of offering Iraq certain facilities on Kuwait Bay, though it was understood that if these were to be offered they would almost certainly have to be at a British-controlled port. With the British undecided on the issue, Colonel Sir John Ward, the British head of the Basra Port Directorate, which administered the Shatt al-Arab waterway from Basra down to the Gulf, suggested that Iraq should be persuaded instead to erect port facilities on the Khor Zubair.[3] The economic potential of the Khor Zubair had been recognised in Ottoman times as early as the mid-1860s. Indeed, it was largely the government of India's fear in the first decade of the twentieth century that an Ottoman/German railway terminus might be developed on the Khor Zubair, that prompted Britain to support the Ruler of Kuwait's claim to Bubiyan island and to encourage him to claim Warba island, lying further north (Schofield 1991). Anyway, by the autumn of 1938, Britain had basically decided, after Ward's suggestions, that the Iraqi government should focus its attention on the Khor Zubair if it wanted to develop port facilities outside the Shatt al-Arab.

Iraqi Foreign Minister Taufiq al-Suwaidi, in an *aide-mémoire* of September 1938, gave three principal reasons why Iraq wanted a new outlet to the Gulf and preferably a position on its coastline.[4] Firstly, the reliability of the Shatt al-Arab was vulnerable to the long-standing dispute with Iran over its international status. After frequent incidents in the Shatt al-Arab during the early 1930s and an unsuccessful reference of the dispute to the League of Nations in late 1934, Iran and Iraq had finally agreed in the Tehran Treaty of July 1937 to confirm for the most part Iraqi sovereignty over the river. The boundary was extended to the *thalweg* opposite the burgeoning Iranian oil port of Abadan in the agreement, much in the way that the boundary had been extended to the median line of the river opposite the Iranian port of Muhammarah (Khorramshahr) in the 1913 Constantinople Protocol. This agreement had defined the whole length of the Perso-Ottoman boundary from Mount Ararat in the north to the Persian Gulf in the south. Yet within months of the 1937 treaty's signature, Iran was claiming that the boundary should have run along the *thalweg* for its entire course along the Shatt al-Arab, a positional demand which was not satisfied until the conclusion of the Algiers Accord in March 1975 and a series of subsequent bilateral boundary treaties signed in June and December the same year.

The Iraqi Foreign Minister's second and more contentious concern was that the Shatt al-Arab was becoming increasingly congested, due principally to the activities of the Anglo-Iranian Oil Company at Abadan. Thirdly, Iraq wished to extend its railway beyond Basra, lying 72 miles upstream from the mouth of the Shatt al-Arab at Faw, to the coast.

In discussions which followed in London between Foreign Office officials and the Iraqi Foreign Minister during October 1938, both sides seemed to accept that in order to develop an Iraqi port effectively at Umm Qasr on the Khor Zubair, certain territorial concessions would be required of Kuwait. The Foreign Office argued that if Iraq wanted Kuwait to cede sovereignty over Warba and over the southern waters of the Khor Abdullah, then it would probably have to offer compensation to the Sheikh of Kuwait which considerably exceeded the actual value of this territory (Schofield 1991). During the summer of 1939, Ward undertook surveys of the Khor Zubair to find a suitable site for the proposed port development. He opted for the same natural anchorage south of Umm Qasr that the government of India had been scared the Germans would develop some three and a half decades earlier. Impressed with Ward's surveys, the Iraqi government announced in November 1939 that wharves and moorings for two ships would be erected at the site and that for the effective protection of the port and its approaches, Iraq would require Kuwait to cede the islands of Warba and Bubiyan. As these islands were barren, occasionally sandy mud flats and of no real value to Kuwait, the Iraqi Foreign Ministry argued that Kuwait need receive no compensation for their cession. The British Ambassador in Baghdad immediately responded that there must be a quid pro quo, and if this could not be territorial then it should be financial.[5]

The British government now turned their attention to whether Iraq should be allowed to secure control over the Khor Abdullah and, if it should, what inducements might be given to the Ruler of Kuwait. Back in August 1939 a Foreign Office official, Lacy Baggallay, had commented that complete Iraqi control of the water inlet would be most convenient for practical administration. During December of that year, evidently backed by the Foreign Secretary Lord Halifax, he added that there would be some justice in this arrangement:

it is understandable that the State which controls the Mesopotamian plain should desire to have undivided control of at least one good means of access to the sea, and Lord Halifax thinks that on a long view it is likely that, if Iraq were given this access, it would make for steadier conditions in that part of the world in years to come.[6]

History has proven Halifax's contention an insightful one. Even then Britain was seemingly aware of the restlessness likely to be engendered by Iraq's disadvantageous position on the Gulf.

The Foreign Office also argued (in the same despatch) that Kuwait need not necessarily give up sovereignty over Bubiyan for Iraq to control the whole of the Khor Abdullah – that is, the boundary might run along the low-water mark of the northern shore of the island. It might be added, though, that 100 years of dispute between the Ottoman Empire and Persia and their successors over a similarly arranged Shatt al-Arab boundary hardly augured well for the scheme's workability.

However, the Foreign Office were quick to stress that whatever Iraq's desiderata, the Sheikh of Kuwait was under no obligation to cede to Iraq any part of his territory, except in return for what he considered to be adequate compensation. This was to be vital, for despite hints during February 1940 from Iraq that economic satisfaction would be given to Kuwait if it agreed to cede the islands,[7] the Ruler of Kuwait, strongly supported by the British authorities in the Gulf (still at this stage employees of the government of India), ruled out any possibility of ceding territory to Iraq in March 1940.[8] The evident tenacity with which Sheikh Ahmad of Kuwait defended every inch of his territory persuaded the British government that there could be no question of Kuwait ceding Warba and Bubiyan to Iraq. So Britain now considered that any basis for the demarcation of the Kuwait–Iraq boundary offered to the Iraqi government had to be strictly in accordance with the express terms of the 1923 and 1932 diplomatic correspondence which had originally defined the boundary, albeit not very clearly. These defining diplomatic exchanges had, however, clearly specified Kuwaiti ownership of the islands of Warba and Bubiyan (Finnie 1992). It was now obvious that Britain would not support any moves for Iraq to gain control of the islands against the will of the Kuwaiti ruler.

As a consequence the British government presented a basis for the demarcation of the Iraq–Kuwait boundary to the Baghdad government during October 1940, which was strictly in keeping with but a detailed elaboration of the 1932 correspondence which had defined the boundary (Schofield 1991: 86–7). Iraq replied that it would strongly deprecate any demarcation of the land boundary before the islands of Warba and Bubiyan had first been ceded to Iraq.[9] So the border and islands question had been set in the pattern of deadlock that persisted right the way up to the Iraqi invasion of Kuwait on 2 August 1990. The upshot of all this was that the Iraqi port on the Khor Zubair was not developed during the 1940s (though an allied port was constructed during the Second World War slightly to the south of Umm Qasr, later dismantled in part because of its sensitive location at the end of this conflict) and Iraq had to continue to rely on the Shatt al-Arab for access to the Gulf.

THE SHATT AL-ARAB CRISIS OF 1969 AND RESULTANT IRAQI PRESSURE FOR KUWAITI CONCESSIONS OVER WARBA AND BUBIYAN, 1969–77

Iran–Iraq relations deteriorated seriously during the first few months of 1969. Shah Muhammad Reza Pahlavi had made no secret of Iran's desire to play a more assertive role in the Gulf since Britain announced its intention to end its long-standing protection of the Arab sheikhdoms along its south-western littoral during January 1968. At the same time, the Iranian government intensified its calls for a *thalweg* boundary de- limitation along the Shatt al-Arab, sovereignty over which was still vested predominantly in Iraq. The Ba´athist government in Baghdad responded by issuing vague claims over the south-western, originally Arab province of Khuzistan (Schofield 1986, Akhtar 1969). Matters then came to a head along the Shatt al-Arab itself. During mid-April the Iraqi government announced that all Iranian ships in the estuary would have to lower their flags, while all Iranian nationals on board ships in Iraqi waters would have to disembark. The Shah immediately responded by unilaterally abrogating the 1937 Tehran Treaty and placing his considerable air and naval forces on high alert. It was further stated that all Iranian and Iran-bound shipping in the Shatt al-Arab would now receive a military escort. Under such conditions Iraq took no actions to ensure that its newly announced regulations were observed by Iranian shipping. With the status of the Shatt al-Arab more insecure than ever, Iraq turned its gaze south once more to Kuwait and the Khor Zubair.

During the last week of April 1969, therefore, a high-ranking Iraqi delegation was despatched to Kuwait to advise the Kuwaiti government of an impending Iranian attack upon Iraq, appealing in the name of Arab solidarity to be allowed to station forces on both sides of the un- demarcated Kuwait–Iraq boundary to protect the recently opened port of Umm Qasr (1961). Kuwaiti ministers were to claim that by the time the request had been made, Iraqi troops had already advanced a few miles into Kuwaiti mainland territory south of Umm Qasr. Since Arab opinion would demand that Kuwait gave some support to a fellow Arab state under threat from Iran, the Kuwaiti government tacitly acquiesced in this *fait accompli* (Schofield 1991).

Once in position south of Umm Qasr, the Iraqi forces did not withdraw (and then only partially) until the major improvement in relations between Kuwait and Baghdad during 1977. Iraq maintained that for as long as the Shatt al-Arab dispute remained unresolved the detachment should remain in Kuwaiti territory. Iraq remained genuinely anxious to secure control over the Khor Bubiyan, lying between the islands of Warba and Bubiyan, which was a much deeper and better approach channel to Umm Qasr port than the Khor Shetana north of Warba island, which, as things stood, was shared with Kuwait. May 1972 saw the first visit by an Iraqi Foreign Minister, Murthatha Abdul Baqi, to Kuwait to discuss bilateral relations and present a scheme for the settlement of the border question. Iraq was apparently prepared to recognise a boundary delimitation based on the 1932 and 1963 agreements, provided that Kuwait recognised Warba, Bubiyan and its northern land territories as strategic regions in which Iraq would be allowed to erect military bases and enjoy unimpaired military access. This Iraqi *diktat* was, not surprisingly, rejected by Kuwait (Schofield 1991). Later in the year the Iraqis returned to their demand that Kuwait should cede Warba and Bubiyan. Iraq outlined its plans for the construction of a deep-water oil export terminal on Bubiyan. Kuwait rejected such a proposal, adding that, in any case, the waters surrounding the island were far too shallow to contemplate such a development. After Kuwait had rejected Baghdad's demands for a sizeable financial loan during December 1972, Iraq considerably strengthened its existing troop detachments in Kuwaiti territory, which had remained in position south of Umm Qasr since April 1969 (al-Mayyal 1986: 137, Kelly 1980). During this period Iraq also constructed a road three or four miles into Kuwaiti territory over which it has only relinquished control since the completion by the United Nations of its demarcation of the Iraq–Kuwait land boundary during late 1992.

After further Iraqi demands during early 1973 that Kuwait cede the islands of Warba and Bubiyan and a strip of land territory south of Umm Qasr were rejected by Kuwait, a contingent of Iraqi forces already positioned in Kuwaiti territory (they were around 3,000 in number at this time (*Arab Report and Record* 1973: 130)) attacked the Kuwaiti border post of al-Samta, killing two border guards. The most significant long-term result of the al-Samta episode was its effect on the territorial consciousness of the Kuwaiti government and public alike. From this time onwards Kuwait became visibly defensive about Warba, Bubiyan and its northern territories. For example Kuwaitis began to name shops, ships and businesses after the islands of Warba and Bubiyan and northern border posts such as al-Samta and Abdaly (Schofield 1991: 117).

With the conclusion of the March 1975 Algiers Accord and follow-up bilateral treaties of June and December 1975, Iran and Iraq agreed a *thalweg* boundary delimitation along the Shatt al-Arab river, thereby satisfying a long-standing Iranian positional demand (Kaikobad 1988; Schofield 1986, 1989). The signing of the Accord had certainly been the result of a destabilised and vulnerable Iraq trying to find an accommodation with its more powerful eastern neighbour, Iran. The strong support lent by the Shah during the early 1970s to Mullah Mustafa Barzani, leader of the Kurdish rebellion, was threatening the imminent collapse of Ba'athist authority over an ethnically heterogeneous Iraqi state. In Baghdad's eyes, a solution to the Kurdish problem temporarily had to override all other considerations. Therefore Iran named its price. In return for abandoning material support for the Kurds, Iran secured an equal share of the Shatt al-Arab river. Suspicions that Iraq would not have concluded such an arrangement had a more equal balance of power existed between the two states were largely confirmed during October 1979 when Saddam Hussein, President of Iraq since July of that year, denigrated and denounced the Algiers Accord. By this time the Shah had been toppled by the Islamic Revolution, one of the major consequences of which had been a dramatic destabilisation in relations between Tehran and Baghdad. Saddam Hussein unilaterally abrogated the Algiers Accord in September 1980 as a prelude to his decision to prosecute war against the Islamic Republic, tearing Iraq's copy of the Algiers Accord to pieces with evident fervour before an Iraqi television audience. His perception that the balance of power across the Shatt al-Arab had tilted back in Iraq's favour had persuaded the Iraqi leader that the time was right to restore the Shatt al-Arab to full Iraqi sovereignty (Schofield 1986: 62–5).

Even though the Algiers Accord had been signed on Iran's terms, it had nevertheless placed relations between Iran and Iraq back on a level footing. Thus, Iraq could no longer really claim with any justification that it needed to retain its forces in Kuwaiti territory for the defence of Umm Qasr (Litwak 1981: 31). Yet, by the mid-1970s, Iraq's incentives to secure complete control over the Khor Abdullah and the islands of Bubiyan and Warba had, if anything, grown. For a start, sovereignty over the Shatt al-Arab was shared, rather than vested primarily in Iraq as had previously been the case. More fundamentally, the capacity of the port of Basra had genuinely reached saturation point by the mid-

1970s (as had not been the case during the 1930s, whatever the claims of Colonel Ward), with cargo ships often having to wait for two months at a time within the Shatt al-Arab before unloading. Though plans were drawn up to construct fresh midstream and dockside berthage capacity to relieve some of the congestion at Basra, it was noticeable that nearly all the financial resources set aside for port development in the Iraqi 1975–9 National Development Plan were earmarked for the expansion of Umm Qasr (Schofield 1986: 74). It seemed clear that the large-scale expansion of the Khor Zubair was to go ahead, whether or not an agreement was reached with Kuwait on the lease or cession of Warba and Bubiyan. This naturally begged the question of whether Iraq assumed that agreement would be reached in due course with Kuwait on the islands issue, or whether it could successfully develop the port without any territorial concessions. After the Kuwaiti National Assembly reaffirmed in July 1975 'Kuwait's sovereignty over all its territory within the borders which have been approved in accordance with international and bilateral agreements between Kuwait and its neighbours' the prospects for the cession or lease of the islands seemed more remote than ever (Day 1982: 225). The question of whether or not the Khor Zubair could be developed without territorial concessions from Kuwait was never resolved, since the proposed expansion of Umm Qasr, already running some way behind schedule, became one of the first economic casualties of the Iran–Iraq war (Schofield 1991: 119–21).

THE ISLANDS OF WARBA AND BUBIYAN DURING THE IRAN–IRAQ WAR AND AFTERWARDS, 1980–90

With the onset of the Iran–Iraq war in September 1980, Iraq's ability and incentive to pursue the economic development of the Khor Zubair suddenly disappeared. Iraq's oil production fell from a pre-war figure of 3.4 million barrels per day to 500,000 barrels per day as a result of its deep-sea oil terminals of Khor al-Amaya and Mina al-Bakr being incapacitated during the first week of the war (Schofield 1986: 78). Basra and Umm Qasr ports were soon reported to be closed, while production from Rumaila and Zubair was curtailed as Iraq had no means of exporting crude from its southern fields.

In February 1981 Iraq announced its readiness to demarcate the Kuwait–Iraq land boundary, providing Kuwait would agree to the lease of Warba and Bubiyan, upon which Iraq wished to develop a military base for use in the war with Iran. Kuwait's resistance to the proposal was only reinforced when Hussein Mussavi announced shortly afterwards, during his brief spell as Iranian Foreign Minister, that the emirate would be dragged into the conflict if it leased the islands to Iraq (Nonneman 1986: 107). In November 1984 a high-ranking Kuwaiti delegation was confronted with a similar request for the lease of the islands during its visit to Baghdad. This came to nothing, though not all interested parties were so sure that the meeting had failed. The Speaker of the Iranian Majlis, Ali Akbar Hashemi Rafsanjani, evidently believed that Kuwait had agreed to lease to Iraq the two islands of Warba and Bubiyan. He cautioned Kuwait that, if leased to Iraq, the islands would be attacked by Iran and not necessarily ever returned to Kuwait. The Iranian statement produced from Kuwait an instant denial and an equally resolute determination to defend the islands. In December 1984 it was announced that Kuwaiti military forces as well as rockets and anti-aircraft batteries were in place on the islands of Warba and Bubiyan. With Kuwaiti garrisons in place on the islands of Warba and Bubiyan and Iran displacing Iraqi troops from the Faw peninsula in February 1986, the questions of leasing the islands and demarcating the boundary remained dormant for the remainder of the Iran–Iraq conflict (Schofield 1991: 122–3).

Kuwait's determination physically to assert sovereignty over Bubiyan had earlier been evidenced in 1983 by the speedy construction of an economically useless bridge linking the island with the Kuwaiti mainland at Sabiya, at the northern entrance to Kuwait Bay. The purely symbolic value of the bridge was clear for all to see since the feature possessed no connecting roads. Hence it was jokingly referred to as 'linking nowhere to nowhere' (al-Mayyal 1986: 242–3, Nonneman 1986: 58). During 1984 the Kuwaiti Ministry of Information announced that a further bridge was to be constructed to link Bubiyan with Warba further north, though there were no indications that this had been completed by the end of the 1980s. The Ministry also revealed proposals to develop recreational and research centres, along with fish-canning plants on Bubiyan (Government of Kuwait 1984: 88–90).

Barely a fortnight after Iran's acceptance of United Nations Resolution 598 (terms for an Iran–Iraq cease-fire), an Iraqi Interior Minister arrived in Kuwait in early August 1988, at the emirate's request, for talks on the border question. It is clear that neither this round of consultations nor a further visit to Kuwait in December 1988 by Izzat Ibrahim, Vice-Chairman of the Iraqi Revolutionary Command Council, produced any concurrence on the border and islands question. Though no details were released, it seems reasonable to assume that Kuwait held out for a demarcation of the boundary according to the 1932 and 1963 agreements. Iraq, by the same token, was probably keen to pursue once more the idea of leasing the islands of Warba and Bubiyan, all the more so in view of the unnavigable state of the Shatt al-Arab after eight years of neglect during the Iran–Iraq war (Schofield 1991: 124). Any idea of conceding control over Warba and Bubiyan was now probably anathema to the Kuwaiti government, which had done much physically to assert its sovereignty over the islands during the Iran–Iraq war. In February 1989 the Kuwaiti Prime Minister Sheikh Saad visited Baghdad, reportedly hopeful that Iraq might go a long way towards meeting Kuwaiti desiderata on the border issue in recognition of the massive financial aid the sheikhdom had provided during the Iran–Iraq war. He apparently met with little success, reportedly rejecting an Iraqi proposal that Warba and Bubiyan should be traded in full sovereignty for inland Iraqi areas (Nonneman 1990: 8). Kuwaiti–Iraqi relations cooled noticeably from this point onwards and, apart from fleeting and unsuccessful consultations during the autumn of 1989, no more high-level efforts were made to solve the border and islands question before the Iraqi invasion of Kuwait on 2 August 1990.

Despite the absence of significant dialogue on the border issue between February 1989 and August 1990, it is probable that Iraq's gaze remained focused primarily upon Kuwait's northern territories and islands as its future means of access to the Gulf. At the beginning of this chapter it was mentioned that Saddam Hussein had once again accepted the *thalweg* Shatt al-Arab delimitation originally introduced by the 1975 Algiers Accord on 15 August 1990 – almost a fortnight after his invasion of Kuwait (there has, as yet, been no formal document signed between the two states to follow up Saddam's strategically motivated action). This move surprised many observers. Only in late May 1990 had all Arab league member states (ironically including Kuwait) present at the extraordinary Baghdad Summit given unanimous support to Iraq's claims to sovereignty over the whole of the river (*Iran Focus* 1990: 6).

Yet perhaps there were some clues earlier in the year that Saddam Hussein was about to reconcile himself to sharing sovereignty over the clogged-up Shatt al-Arab waterway with Iran. An exchange of correspondence between the Iraqi and Iranian leaders beginning in April 1990 was hinting at greater Iraqi flexibility towards existing Iranian terms for peace. Even so there is no clear indication that Iraq took the decision to abandon claims to the whole of the Shatt al-Arab river until such time as the decision to invade Kuwait had been made, that is at some time in mid to late July 1990 (Schofield 1993b). Yet with the benefit of hind-sight it might be argued that Saddam Hussein had calculated by the end of 1989 that the only realistic means of improving Iraq's access to Gulf waters was at the expense of Kuwaiti national territory. By this reckon-ing, however, no one could have expected him to take more than Warba and Bubiyan.

CONCLUSION

An examination of the three episodes in the long-running Iraqi demand for the cession or lease of the Kuwaiti islands of Warba and Bubiyan has certainly highlighted the major influence of the Shatt al-Arab dispute and Iranian–Iraqi relations more generally. In the late 1930s Iraq's dissatisfaction with the status and increasing congestion of the Shatt al-Arab was an important factor in the decision to seek alternative port facilities on the Khor Zubair. In 1969 Iraq used the crisis over the Shatt al-Arab to justify the positioning of Iraqi troops within Kuwaiti terri-tory, where they stayed until 1977, and thereafter in smaller numbers. Despite the extensive efforts of the Iraqis in this period to get the Kuwaitis to agree to the cession or lease of the islands of Warba and Bubiyan, Kuwait proved no more flexible than it had done for the previous three decades. At the end of the Iran–Iraq war Kuwait was in no position to give up sovereignty of control over the islands of Warba and Bubiyan. It had done much to bolster sovereignty over these features during the conflict, in part because it felt threatened by the proximity of the fighting and in part because it could do so without fear of much reaction from Iraq, who had more pressing concerns. The names of the islands were now also firmly stamped on the territorial consciousness of the Kuwaiti government and public alike.

In the Gulf War cease-fire resolution (United Nations Resolution 687 of 3 April 1991), the United Nations undertook to guarantee the existing Kuwait–Iraq boundary (as referred to in the 1963 'Agreed Minutes...'). This and the institution of a team selected by the United Nations Secretary-General to demarcate fully this troublesome terri-torial limit (United Nations Iraq–Kuwait Boundary Demarcation Commission (UNIKBDC)) was accepted by both Kuwait and Iraq (albeit grudgingly). The demarcation team's decision on the land boundary was announced in April 1992, an award all but rejected by the Iraqi government, which withdrew its delegate from the activities of the commission some two months later. UNIKBDC announced the de-marcation of the land boundary during November 1992. An announcement on the course of the maritime boundary along the Khor Abdullah is not expected until the spring of 1993, though it is assumed that the delimitation, when announced, will follow the median line and not the *thalweg*. The latter had been employed in previous demarcation proposals presented unavailingly to the Iraqi government during 1940 and 1951. The median line was confirmed as the boundary along the Khor Abdullah by UNIKBDC as expected in March 1993.

Notwithstanding the recent 'successes' of the United Nations, it would take a brave person to say that Iraq–Kuwait territorial disputes have been finally settled. Ominously the UN verdict on the land boundary has been rejected by not only the current Iraqi regime, but by nearly all of those opposition forces which the West would supposedly rather see ruling in Baghdad. Similarly the dispute over the Iran–Iraq boundary along the Shatt al-Arab must be regarded as currently dormant rather than permanently settled. It could be resurrected every bit as quickly as was that other fault-line in Irano-Arab territorial affairs during 1992, the Abu Musa/Tunbs dispute (Schofield 1993a).

In the long term it is ultimately of the utmost importance for the future stability of Kuwaiti–Iraqi and also Iranian–Iraqi relations that Iraq no longer perceives itself as being squeezed out of the Gulf. Given its narrow coastline and the fact that Iraq exercises complete sovereignty over neither the Shatt al-Arab nor the Khor Abdullah, its two means of access to Gulf waters, this will probably prove a very difficult perception to assuage. It remains to be seen whether Kuwait will come under renewed Iraqi pressure to make concessions on the islands issue when Baghdad's relations with Tehran next deteriorate seriously over the status of the boundary along the Shatt al-Arab – traditionally, during the last two decades at any rate, the cue for Iraq to press territorial demands on Kuwait. Unless a satisfactory long-term solution to the problem, genuine or illusory, of Iraqi access to the Gulf can be found, then successive Baghdad governments may well continue to argue that Kuwait should compensate Iraq for its geographic and strategic misfortune.

NOTES

1 For example, the comments reportedly made by the Iraqi Foreign Minister during April 1973: see *Kuwait Times*, 5 April 1973.
2 Despatch dated 30 March 1938 from Sir Maurice Peterson, British Embassy, Baghdad to the Foreign Office; in the India Office file, no. R/15/1/541.
3 Despatch dated 26 August 1938 from Lacy Baggallay, Foreign Office to R.T. Peel, India Office; in the India Office file, no. R/15/5/208.

4 *Aide-mémoire* dated 28 September 1938 by Taufiq al-Suwaidi; in the Public Record Office file, no. FO 371/21858.
5 Despatch dated 14 November 1939 from Houston-Boswell, British Embassy, Baghdad to Lord Halifax, Foreign Office; in the Public Record Office file, no. CO 732/86/17.
6 Despatch dated 16 December 1939 from Lacy Baggallay, Foreign Office to the India Office; in the Public Record Office file, no. CO 732/86/17.
7 Telegram dated 27 February 1940 from Sir Basil Newton, British Embassy, Baghdad to the Foreign Office; in the Public Record Office file, no. CO 732/86/17.
8 Telegram dated 22 March 1940 from the Political Resident in the Persian Gulf to the Secretary of State for India; in the Public Record Office file, no. FO 371/24559.
9 Despatch dated 29 November 1940 from Newton, British Embassy, Baghdad to Lord Halifax, Foreign Office; in the Public Record Office file, no. FO 371/61455.

REFERENCES

Akhtar, S. (1969) 'The Iraqi–Iranian dispute over the Shatt al-Arab', *Pakistan Horizon* 24(3), 213–21.
Al-Mayyal, A. (1986) 'The political boundaries of the state of Kuwait', Unpublished Ph.D. thesis, School of Oriental and African Studies, University of London.
Arab Report and Record, no. 6, 16–31 March 1973.
Day, A.J. (1982) *Border and Territorial Disputes*, London: Longmans/Keesing's Reference Publication.
Finnie, D.H. (1992) *Shifting Lines in the Sand: Kuwait's Elusive Frontier with Iraq*, London: I.B. Tauris.
Government of Kuwait, Ministry of Information (1984) *Kuwait, Facts and Figures*, Kuwait.
Iran Focus, June 1990.
Kaikobad, K.H. (1988) *The Shatt al-Arab Boundary Question: A Legal Reappraisal*, Oxford: Oxford University Press.
Kelly, J.B. (1980) *Arabia, the Gulf and the West*, London: Weidenfeld & Nicolson.
Kuwait Times, 5 April 1973.
Litwak, R. (1981) *Security in the Persian Gulf 2: Sources of Inter-State Conflict*, London: International Institute for Strategic Studies, Gower.
Nonneman, G. (1986) *Iraq, the Gulf States and the War: A Changing Relationship, 1980–1986 and Beyond*, London: Ithaca Press.
—— (1990) 'Iraq and the Arab states of the Gulf: modified continuity into the 1990s', Paper presented at the Royal Institute of International Affairs, 9 May.
Schofield, R.N. (1986) *Evolution of the Shatt al-Arab Boundary Dispute*, Cambridgeshire: Menas Press.
—— (ed.) (1989) *The Iran–Iraq Border, 1840–1958*, Farnham Common: Archive Editions.

—— (1991, 1993b) *Kuwait and Iraq: Historical Claims and Territorial Disputes*, London: Middle East Programme, Royal Institute of International Affairs.

—— (1993a) 'Borders and territoriality in the Gulf and Arabian Peninsula during the twentieth century', in R.N. Schofield (ed.), *The Territorial Foundations of the Gulf States*, London: UCL Press.

11

THE UNITED ARAB EMIRATES AND OMAN FRONTIERS

Julian Walker

FRONTIERS IN ARABIA

The concepts of nation-states, of territorial sovereignty, and of fixed linear frontiers are Western ones, which have been imposed on the traditional society of the Arabian Peninsula. In that area the true 'Arab' bedouin tribes have been accustomed to roam for hundreds of miles between their winter and their summer grazing. The men and their camels mattered, rather than the comparatively barren land, and they switched their allegiance from one leader to another as their relationship with him, and their hopes of gain as a result of their adherence to him, changed. As a result the power of the leading families of the area fluctuated, so did the extent of their territorial influence. It was the wells and the grazing which gave the desert terrain the little importance that it had. Only in areas where settlement and agriculture (themselves despised by the warrior of the desert) were possible, did the land have value. Thus it was in such areas, as in the mountains of Yemen and Oman, that the idea of a 'people' with a common history and tradition emerged, to create a concept which had some similarities with the idea of the 'nation' which existed in the West.

As a result the first frontiers in the Arabian Peninsula arose not from indigenous demands but from the need of powers from outside the area, the British and Ottoman Empires, to delimit their spheres of influence where these came into contact with each other. Thus the frontiers between the two Yemens were partially delimited, the Ottoman sphere of interest in the Rub' al-Khali restricted by the Blue and Violet lines, and the extent of the Sheikhdom of Kuwait agreed upon before the First World War. After that war there was a pause in the creation of frontiers until the rise of Ibn Saud's Najdi Kingdom, which, together with the rise of British and French mandates to the north, made further line-drawing advisable. Even then the Protocol of Uqair in 1922 was not meant to restrict the wanderings of the great *bedu* tribes, the Dulaim, Ruwalla, Anaizah, Shammar and Ajman, and neutral zones were created to give certain tribes the right to common grazing. The continuing expansion of Saudi rule gave rise to further frontier drawing from the late 1920s until the early 1930s when the Treaty of Taif resulted in the demarcation of a border between Asir and the Yemen, the first frontier in the Peninsula fixed by Arabs rather than outsiders.

THE DISCOVERY OF OIL

With the discovery of oil in Bahrain in 1932, followed by that in the Hasa, the Western world began to impinge more deeply on the life of the Peninsula. The scramble for oil concessions that followed coincided with a period of Saudi/British negotiations which attempted to settle the boundaries round the periphery of the Empty Quarter, where Saudi influence met that of the rulers of the British-protected states bordering the coast between Jabal Nakhsh in Qatar in the north and Aden in the south. These negotiations, while they succeeded in laying down general parameters for future frontiers between the Fuad Hamza and Riyadh lines, ceased before any agreement was achieved, partly because of the outbreak of the Second World War. The Hamza line, offered to the British government by Deputy Foreign Minister Fuad Bey Hamza during April 1935, remained Saudi Arabia's claim to territory in southern and south-eastern Arabia until the extension of October 1949, when the Buraimi oasis was claimed. The Riyadh line remained, for the course of the intermittent Anglo-Saudi frontier negotiations, Britain's most generous offer to Saudi Arabia for territorial limits in the region. While the Hamza line was quickly rejected by Britain, the Saudi government rejected the Riyadh line within 24 hours. The failure to reach agreement then laid up trouble for the future, and in 1949 tension between Saudi Arabia and the coastal states led first to the Buraimi dispute and, thereafter, to troubles in Inner Oman, which ended with the capture of the Jabal Akhdar in January 1959. Until these troubles attracted attention to them, the territories in the interior of the Trucial States and the Sultanate of Oman, had been a tribal backwater, hardly visited by the British since the beginning of the century. In Oman the Agreement of Sib in 1920, and the financial difficulties of the Sultan of Muscat, had prevented the latter from consolidating his suzerainty over the tribes of the interior. In the Trucial States the collapse of the pearl trade in the early 1930s had left the sheikhs of the coast desperately poor and with little control of the tribes and the slave traders roaming inland. The Abu Dhabi/Dubai war of the late 1940s had added to the lack of security in the area, even if, at the end of it, the British had imposed a stretch of frontier between the two sheikhdoms, running south eastwards from the coast at Ras Hasian.

The need to counter the Saudi penetration of Buraimi and beyond changed the situation radically. The Trucial Oman Levies, created in 1951 to combat slave trading and protect British officials and oil company parties, had to be rapidly expanded and despatched to Buraimi to support Sheikh Zaid of Abu Dhabi in blockading the Saudi representative, Turki bin Ataishan, in Hamasa. The Sultan, having originally summoned the Imam of Inner Oman to advance with his forces to expel the intruders from Hamasa, was dissuaded from attack by the British, and used some of his men to complete the blockade. Later, other Sultanate forces escorted a Petroleum Development Oman (PDO) oil prospecting party from the southern coast of the Sultanate to Fahud in the tribal territory of the Duru, and then were persuaded to take the Duru's market town of Ibri, about 100 miles south of Buraimi. As a result the interior of the Sultanate, which had not been visited by the British since the nineteenth century, was opened up to oil prospecting.

INTER TRUCIAL STATES FRONTIERS

Oil prospecting on the Trucial Coast made the local oil company, Petroleum Development Trucial Coast (PDTC), keen that the boundaries between the various sheikhdoms there should be defined. The rulers themselves, anxious to encourage exploration by the oil company in the hopes that it could foreshadow the end of the poverty which had affected the area since the collapse of the pearl trade, and wishing to reduce friction between them, were, with one exception, happy to accept the proposal of the British Political Agent at Dubai, Christopher Pirie Gordon, that he should arbitrate over their frontiers. They agreed that they would not dispute the results of his arbitration. The exception was Sheikh Shakhbut of Abu Dhabi, who had been incensed by the award made by Pirie Gordon's predecessor, Pat Stobart, of the frontier line at Ras Hasian between himself and Dubai, and who was happily aware that the oil company were committed to drill for oil in his territory to the west of Abu Dhabi town, miles away from any possible territorial claims from his brethren of the northern Trucial Coast.

Christopher started the arbitration work by touring the frontier between the two sheikhdoms of Umm al-Qaiwain and Ras al-Khaimah, about which there was said to be no dispute. But he was sadly disillusioned by his first day's experience in the field, and, as a result, decided that it would not be possible for him to carry out the work necessary for frontier arbitration and also pursue his normal office work. He therefore decided to depute the frontier work to his assistant in the Political Agency, and the task fell to me.

As a result I started, in the winter of 1954/5, gathering the information needed for the settlement of the frontiers between the rulers of the northern Trucial Coast. We were especially interested in identifying the frontier points on the coast itself, as PDTC (Petroleum Development Trucial Coast) had just lost their case as holders of the mainland concession areas for claiming that the Gulf sea-bed was allocated to them under the concessions that they had signed with the local rulers in 1937, and the latter therefore hoped to conclude new concession agreements covering the sea-bed. It soon became obvious to me, given the nature of the area, that there was no possibility of frontier settlement other than on the basis of the tribes and local history. Much of the latter could be gleaned from the files of the Political Residency in Bahrain and of the Agency in Dubai, including Lorimer's invaluable *Gazetteer* of 1915 which named villages and tribes which had disappeared from our knowledge since the time of its issue. These records could be supplemented by the writings of local historians, and the memories of the older *bedu*, who could recount to me incidents with such clarity that they might have happened the week before. The history revealed past acts of influence and control, such as the cementing of wells, and the pursuit of murderers and camel raiders. There was also more contemporary evidence, such as the collection of *zakat* on crops and livestock, the control of *sakham*, charcoal production for commercial purposes, and *naub* – the organisation for a fee of the supply and distribution of irrigation water. Tribesmen living in the disputed areas could also give valuable evidence of control by rulers, as long as their fears that what they said might leak to those whose cause it damaged, and result in their exposure to retribution from angry claimants, could be overcome. But it was not only evidence that I had to gather. There were no accurate maps of the area, and in some cases no maps at all, so these had to be made at the same time as I gathered evidence indicating where a frontier might lie. I would tour the area of frontier claimed by one ruler with his adherents, and then do the same with the adherents of his opponent, at the same time trying to map the whole of the territory which seemed to be in dispute. If the area between the two claims was too wide I would have to fill in the gaps on my own, with the help of any *bedu* that I happened to meet on my way.

By the spring of 1955 we were ready to announce our first decisions about frontiers, covering many of the points on the coast and a stretch of border between the sheikhdoms of Ras al-Khaimah and Umm al-Qaiwain, as well as general statements of recognition that certain areas were under the sovereignty of one ruler or another. However, the legal advisers in the Foreign Office were worried that the publication of the criteria that we had used in coming to our conclusions might well be used against us by the Saudis and their American oil concessionaire Aramco, which was helping them to prepare their case for the arbitration of the Buraimi dispute in order to strengthen their case against the Ruler of Abu Dhabi and the Sultan of Muscat. We were therefore told to delay the announcement of our awards, and I was asked to write a confidential report on my findings and the evidence that I had unearthed to support them. I completed this task by the end of May 1955 and went on leave.

In October 1955 the Buraimi arbitration collapsed. The Trucial Oman Scouts and the Sultan's forces reoccupied the oasis and the British announced, unilaterally, a frontier between Saudi Arabia on the one hand, and Abu Dhabi, Muscat and Oman and the Aden Protectorate on the other which they would observe themselves. Having returned off leave I found myself re-posted from Dubai to a less interesting job in the Residency in Bahrain, and soon found myself having to do routine administration work there. That was hardly to my taste, and I therefore suggested that, now that the Buraimi dispute was behind us, our local frontier settlement on the Trucial Coast might go ahead. It took some time to convince the Foreign Office of this, but by March 1956 I was released to go back to Dubai. Thereafter I spent a pleasant six weeks on the coast, making the comparatively easy frontier decisions that I had thought possible the year before.

My original success in getting released from administration work now had its penalties. The British Political Resident, Sir Bernard Burrows, had been impressed by the ease with which I had made some comparatively non-contentious frontier awards and had become enthusiastic about the Trucial Coast frontier settlement. PDTC was committed by their concession agreement to drill a well in Sharjah territory by the end of the year, and the area that they were interested in lay comparatively close to Dubai terrain. The Sharjah–Dubai border therefore needed delimiting that summer, and I was the obvious person for the job. As a result I found myself back on the coast in the sweltering heat of June to tackle a frontier dispute exacerbated by age-long rivalry between the two sheikhdoms. I considered myself lucky when I managed to run a line between the two sheikhdoms running some 50 miles inland. The rivalry over certain wells, like Tawi Bida'at and Tawi Tai, was so heated that I was compelled to leave unsettled areas until I could make further investigations to discover where the frontier should run.

I returned wearily to the Residency to discover that Sir Bernard Burrows was now enthusiastically set on my completing the task of the frontier settlement on the coast the following winter. This included tackling areas which I knew were the most difficult, including the mountain tracts in dispute between the rulers of Ras al-Khaimah and Fujairah, where it was obvious that both states had much to support their claims. In the end I found it necessary to award sovereignty to the Ruler of Fujairah, whose control of the tribesmen living there meant that he could prevent anyone else entering the mountainous tracts without his agreement. But I also had to create a form of shared rights in the territory, to allow the Ruler of Ras al-Khaimah to continue to collect the *zakat* which he had traditionally levied, and to take one-third of any profits from any potential future exploitation of minerals.

THE AWKWARD OUTCOME

The awards between the rulers of the northern Trucial coast resulted in a patchwork quilt of territory – both illogical and impossible from the neat Western viewpoint. Sharjah was divided into six enclaves, tiny Ajman into three, and Dubai and Ras al-Khaimah into two each. Only Umm al-Qaiwain had a single bloc of territory belonging to the al-Ali tribe. Tribes were sometimes split between two or more sheikhdoms, and where the claims of the Sultan of Muscat impinged on the Trucial Coast any settlement had to be suspended. Omani land lay to the north of the Trucial States where the Shihuh tribe inhabited the towering peaks of the Ruus al-Jibal. There was an island of Omani territory in the Wadi Madha, in the centre of the eastern coastal district of Shimailiya, where once again the Shihuh resisted the claims of the Qawasim Rulers of Sharjah and Ras al-Khaimah. Omani territory also lay astride the main route between the eastern and western coasts of the Trucial States, straddling both ends of the Wadi Qor. However, the settlements were, as intended, a confirmation of the situation as it existed on the ground, designed to alter that situation as little as possible. It would have been impossible to get the rulers to live with any other type of settlement. As it was, although the process took long hours of argument and persuasion, and although there was continued protest over certain particularly sensitive points, the awards were, as a whole, accepted by the rulers. They resulted in a reduction of the friction which was inevitable between them. Luckily the rulers all knew each other well and understood that their territories were so intertwined that it was essential that there should be cooperation between them.

TRUCIAL STATES–OMAN FRONTIERS

Cooperation between the Trucial Coast and the Sultanate of Muscat was not so developed, and minor skirmishing around the Wadi Madha became the subject of increasingly heated telegraphic correspondence between the British Residency in Bahrain and the Consulate General in Muscat. In the end Sir Bernard Burrows decided to send me down to Muscat to call on the Sultan with the Consul-General, and to prove to him that we British too had gained some understanding of the complexities of Arabian territorial claims. The result of several hours of discussion with the Sultan was that he invited me to come the following winter to tour his frontiers with a member of the Sultanic Ministry of the Interior, Sheikh Sakhr bin Hamad al-Maamari. As a result I found myself, the following January, setting up camp in Buraimi and, together with the sheikh, whose long white beard and white turban were a distinguishing mark of our mixed company, touring the areas claimed by the Omani and the Trucial Coast tribes to the north of the Buraimi oasis. The Sultan was originally insistent that we should not go south of the Oasis into the Dhahira, where his tribes were still inclined to press their claims to independence. Even to the north of the Oasis there were some difficulties, since the Qawasim rulers of Sharjah and Ras al-Khaimah fell heir to a long-standing rivalry with the Sultanate, and the former nursed claims from the distant past stretching as far south as Dhank, forty miles beyond Buraimi. More recently he had exercised control over the northern Shuwaihiyin section of the Bani Kaab tribe. In the end the Ruler of Sharjah decided that he did not wish to seek any frontier settlement with Oman, and the Ruler of Ras al-Khaimah, partly out of solidarity with his brother Qasimi, and partly because he had been stung by the loss of mountain areas to Sheikh Muhammad of Fujairah, also declined to negotiate, through me, with the Sultan.

But during that first winter there was no mention of frontier settlement as the objective of our work. I needed, first, to gain the confidence of the Sultan to tour the frontier areas lying to the south of Buraimi as well as those to the north. It was already warm Ramadhan weather when we were finally allowed to visit the Al Bu Shamis areas to the south of the Oasis, and by that time Sheikh Sakhr bin Hamad, many years my senior, was becoming increasingly disenchanted with the bouncing and dusty progress made by our Landrover. The incoherent guidance of a fasting tribesman in the desert to the south-west of Buraimi finally exhausted his patience and led him to believe that our progress would end with whitened skeletons in the sandy wastes. So he opted out, and by the end of the season the Sultan agreed that I should continue the next year alone, and see whether there were areas where the tribes might agree on dividing lines between them. Thus we sidled towards frontier settlement by mediation. But progress was very slow, and I could sense the impatience of the Foreign Office the following winter as no agreements seemed to be forthcoming. It was a great relief to me when the Ruler of Fujairah accepted a four-mile stretch of frontier to the east of the Wadi Qor which at last left me with something to show for my labours. Thereafter the agreed lines gradually began to appear. To the north of Buraimi Sheikh Zaid of Abu Dhabi and Sheikh Abdullah bin Salim of the Bani Kaab cooperated and compromised to achieve a settlement. In the Oasis itself Sheikh Zaid outlined a claim which was so moderate as to be acceptable to the Sultan. Further south the Al Bu Shamis tribe and the Naim both managed to negotiate agreement on the division of their territories with Sheikh Zaid, with whom they had cooperated against the Saudis.

Time ran out that season, and I had to return for a further winter to tackle other frontiers. The Duru and the Awamir tribes far to the south of Buraimi were feuding with each other and a truce had to be arranged. Even after this had been done the Duru sheikh made such extensive territorial claims that the Sultan of Muscat had to override him in order to get an agreement on the border. However by the end of the two winter seasons we had an agreed line running between the Ramlat Anaij to the north of Buraimi, through it and some 120 miles southwards to a point four miles north of Umm as-Zamul, in the Empty Quarter. This was the Awamir raiding well which had been taken by the British in October 1955 as the point where the frontier with Saudi Arabia met those with Abu Dhabi and the Sultanate. The Sultan, while accepting that Umm as-Zamul most probably belonged to Abu Dhabi as an Awamir well, would not abandon a claim which he had made publicly. Elsewhere we had an agreed frontier between Ajman and Dubai on the one hand, and the Sultanate on the other, to the south of the Wadi

Hatta, straddling the Hajar mountain range that formed the spine of the eastern Trucial States and the Sultanate. I had experienced difficulties between Ajman and the Bani Kaab at the eastern end of this portion of the line, in the Wadi Hadf near Masfut, where the contending sheikhs met face to face in hot discussion over territory where they had conflicting rights. We had to post TOS (Trucial Oman Scouts) forces to keep the rivals apart. In the end the Wadi Hadf became a small neutral zone, shared between Ajman and Oman. Further to the east and north, the Ruler of Fujairah had agreed his four-mile line at Wahala, east of the Wadi Qor. On the southern edge of the Ruus al-Jibal the friendship between his tribe, the Sharqiyin, and the Shihuh had meant that they could agree to the description of a boundary to the north of Dibba without my having to climb the precipitous mountain slopes.

THE AWKWARD RESULTS AND THEIR SURVIVAL

Thus, by the spring of 1960 there was a punctuated frontier, with no agreement between the Qawasim of Sharjah and Ras al-Khaimah and the Sultanate. The precipitous nature of the mountains lying between Ras al-Khaimah and the Ruus al-Jibal would have made settlement in the area difficult even if the relations between the Shihuh and Sheikh Saqr bin Muhammad had been friendly. The Wadi Madha had proved too sensitive to touch. In the area to the south of Kalba at Khatmar Malaha the Sultan stuck to claims that he had made earlier and considered he could not withdraw. In the Gharif Plain to the west of the Wadi Qor and the Hajar Range, Sharjah and the Bani Kaab maintained their dispute. But both in that area and in the mountains of the Shihuh I had some idea of where a reasonable frontier might run. Elsewhere the agreements were signed between the rulers and the Sultan, illustrated by my sketch maps, and accepted by HMG (Her Majesty's Government).

Were the frontiers suitable and acceptable? Once again they were very awkward to Western eyes. Oman was split into three parts by the territories of the Trucial States. The routes between Sultanate territory in Buraimi and Muscat, and from Buraimi to Ibri, ran across land belonging to Abu Dhabi – though it would be possible to build new roads which would avoid it. The best route between Abu Dhabi territories at al-Ain and Umm as-Zamul lay across gravel plains that belonged to Oman. However, once again the agreements represented the situation as it existed on the ground. With both the internal frontiers of the Trucial Coast and those with the Sultanate there were some immediate points of friction. The Ruler of Fujairah manoeuvred against the rights of the Ruler of Ras al-Khaimah in the mountain area over which he had been given sovereignty. Where the territory of Fujairah met with that of Sharjah on the east coast, and rights of property and sovereignty conflicted, the situation was delicate, and a TOS officer was given the task of doing his best to sort the details out. Later, when the Trucial Oman Scouts lost their sense of purpose after the Emirates gained their independence from Britain in 1971, fighting flared between Fujairah and Sharjah and the Abu Dhabi navy arrived in the nick of time to prevent the town of Khor al-Fakkan falling to Fujairah. As I have mentioned, there were still gaps in both the internal and the external frontiers. I also made mistakes. For instance, I mapped Tawi Awaina to the east of Buraimi, further north than it was. There were several natural markers mentioned in the frontier agreements which my successors found it difficult to trace. Dubai and Sharjah persisted in feuding over the coastal point between them at al-Mamzer, and only reached agreement on a modification a few years ago. The wells at Tawi Tai and Bidaat continued to cause difficulties for them, and I believe that they have had to reach a more detailed agreement further south where the al-Maghram gas field was found to straddle the border. Now I have heard that there is difficulty over the frontier which was originally the easiest to settle, that between Ras al-Khaimah and Umm al-Qaiwain. I did not tour that frontier properly myself, and may now have to give some advice about it.

As a whole the internal frontiers have, with minor changes agreed by a committee established by the UAE, held and proved useful. This has basically been because of cooperation between the neighbouring sheikhdoms, and such cooperation has become a habit between them, rather than because the frontiers themselves were designed to avoid difficulties because of inconvenience. The creation of the UAE itself was a sign of this continuing cooperation, and ensured that those frontiers, never meant to be insuperable customs barriers between the neighbouring sheikhdoms, would have little divisive effect on the people living on either side of them. However new routes have been created to link the Union so that the route between the east and west coasts no longer passes through Omani territory on either side of the Wadi Qor, but runs down the Wadi Ham. Even Abu Dhabi and Dubai have agreed on the modification of the line between them, and the customs post on this line has at last been removed. The Qawasim have come to terms with the Sultanate and agreed on frontiers between them. There are still awkward points, but the Sultanate of Oman and the Trucial States have probably been able to relate to each other in a way that would not have developed if there had been no agreed frontiers between them and territorial disputes had proliferated. As a whole I believe that it is not the practicality of the frontiers which has made them a success, if such they are, but the reasonableness of the peoples that they help to unite, rather than divide. In any case I suspect that everyone here will agree that, whatever the attitudes of the local inhabitants, it is better to have frontiers than be without them. Without them we would have no conference and the inhabitants little peace.

12

CAPTAIN KELLY AND THE SUDAN–UGANDA BOUNDARY COMMISSION OF 1913[1]

Gerald H. Blake

INTRODUCTION

The final two decades of the nineteenth century and the first two decades of this century might be described as the formative years of the modern political map of the world. Imperial boundary making led by the European powers was at its height and Britain and France in particular were busy carving out their overseas possessions. In the modern world boundary delimitation has become a lengthy and costly process often involving teams of experts making use of masses of evidence. The imperial boundary makers were not so careful, and often boundaries were drawn up in a hurry with little or no regard for the underlying physical and human geography. Thus a few powerful individuals were able to play a crucial role in the delimitation of boundaries on behalf of their governments. The work of a number of British imperial boundary makers is well documented and might provide rewarding material for further research. One of the most extraordinary testimonies was that of Sir Claude MacDonald who described the early delimitation of the Nigeria–Cameroon boundary (1889–90) to a meeting of the Royal Geographical Society on 9 March 1914 thus:

> In those days (1889) we just took a blue pencil and a rule, and we put it down at Old Calabar, and drew that blue line up to Yola, and that is the boundary ... The following year I was sent to Berlin to endeavour to get from the German authorities some sort of modification or rectification of the blue line, and the instructions which I received on that occasion ... were ... to grab as much as I could. I was also provided at that time with the only

Plate 12.1 Captain Harry Kelly, RE, killed in action in France,
24 October 1914

map – the same map on which we had drawn the blue line. That was nothing more or less than a naval chart! It had all the soundings of the sea very carefully marked out, but the whole of the rest of the sheet was white! There was certainly one thing there, and that was a beautiful river called the Akpayaff, which started near the Calabar river and meandered for about 800 miles on the map. It was about the size of the Amazon, and the idea was that that was to be the boundary – the Germans one side and the

English the other. When we came to close quarters with the Akpayaff river we found there was no such river. There was a river, but so far from being 700 miles long it was only about 3½.

(Nugent 1914: 630–51)

The 360-mile Nigeria–Cameroon boundary was properly demarcated by an Anglo-German Commission of 1913–14. At about the same time Britain had decided to establish a commission to delimit the Sudan–Uganda boundary. The commission assembled at Nimule on 15 January 1913. The senior commissioner was Captain Harry Kelly, who was assisted on the Sudanese side by Captain Bruce, while Uganda was represented by Captain Tufnell and Captain Lilley. Captain Kelly kept a diary throughout the period of the commission and used it to compile regular reports to Wingate, Governor-General of Sudan. The diary is now in the Sudan Archive at the University of Durham, together with Kelly's photographs of the commission. It provides important evidence about the problems of African boundary delimitation in those days, the attitude of the commissioners to their work, and above all the personal relationships between the commissioners themselves, which was to have a fundamental effect on their work.

POLITICAL BACKGROUND

Britain and Egypt had established a condominium over Sudan in 1899. At much the same time the British protectorate of Uganda was gradually extending its influence northwards, and in 1902 the northern limits of Uganda were indicated in general terms by Sir Harry Johnston along latitude 5° North (Figure 12.1). In 1894 Britain and Belgium had delimited their spheres of influence in the region of the Nile–Congo watershed, part of which, in a region known as 'the Lado enclave' was leased to Belgium for the lifetime of King Leopold. When he died in 1909 the transfer of the Lado enclave to Sudan created a tract of Sudanese territory which would clearly be more easily managed from Uganda. It was therefore decided to transfer the southern part of the Lado enclave to Uganda in return for a portion of northern Uganda which would become part of Sudan. In very broad terms the boundary would run from the Lado enclave to the northern end of Lake Rudolf. The Boundary Commission was charged with making recommendations for the alignment of the new boundary east of the Nile. West of the Nile the former Anglo-Belgian boundary had been broadly established, but not properly delimited. It was not apparently included in the commission's terms of reference, although proposals for the line west of

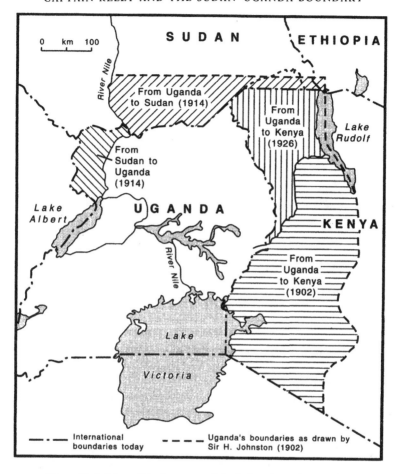

Figure 12.1 The political context of the Sudan–Uganda Boundary
Commission in 1913

the Nile were made by the commission. One stipulation made by the
Foreign and Colonial Offices was that as far as possible tribal boun-
daries were to be respected and tribes were not to be divided between
Sudan and Uganda.

The tract of territory in question had not yet become subject to
effective administration. It was known as dangerous territory where
ivory hunters, bandits, and quarrelsome tribes would not take kindly to
the imposition of law and order. Accordingly, the Boundary Commission set out with an escort of 170 troops (30 Sudanese Camel Corps,

Plate 12.2 Members of the Boundary Commission with District Commissioners at Nimule, January 1913; Kelly is seated second from left, Tufnell on the right

100 infantry of the 12th Sudanese Regiment and 40 men of the King's African Rifles). They were supported by a half section of field hospital, three veterinary officers, and half a dozen Egyptian and Sudanese officers. To carry the equipment and provisions for this large number of men there were 50 camels, 320 donkeys, and 20 mules. Such a large force was probably fully justified. There had been persistent reports during 1910–12 of tribal fighting and gun-running in northern Uganda. Alarming reports had reached Nimule that the Acholi people had at least 2,000 rifles among them, including breech-loaders. In north-eastern Uganda, in the Karamoja and Turkana districts, there were reports of substantial invasions of armed groups from Ethiopia, alleg-edly in an effort to establish Ethiopian territorial claims.

THE RECORD OF THE DIARY: 15 JANUARY TO 16 FEBRUARY 1913

The first sector of the boundary as far as Madial was delimited over some 32 days. The length of the boundary is about 120 miles although the distances covered by the commission in their investigations were far greater. Kelly's faithful servant Adam recorded distances of 8 to 12

Plate 12.3 Kelly described these men as his 'intelligence staff', including Adam
with the perambulator wheel

miles each day with his perambulator wheel. Kelly himself was often up
early in the morning to take bearings from the nearest high ground, so
he probably covered even greater distances.

The chief interest of the diary in this sector is the genuine attempt to
draw up a boundary which would not divide tribes. The difficulties of
achieving this were to prove almost insurmountable, not least because
of the complexity of African tribal relationships which neither Kelly nor
Tufnell were qualified to understand. The evidence of local people does
not appear to have helped very much. Kelly's diary on 1 February is
typical of the frustration he felt:

> half a mile from Khor Koss; still many palms everywhere. The two
> guides from Lokidi proved quite intelligent, but the information
> which I gained from them regarding the various hill people
> make[s] the question of frontiers more complex than ever. The
> Imatong people talk a language resembling Latuka and have been
> to Tarangole to salute Lokidi, but they mostly stay on their hills
> and are not by any means pure Latuka – probably an Acholi
> mixture.

189

In spite of these worthy attempts to fix the boundary with regard for tribal territories, it does not always appear to have outweighed adminis- trative convenience. Thus the Madi tribe were knowingly divided by the Sudan enclave south of Nimule for reasons explained on 20 February: 'Finally settled with Tufnell the Madi frontier line which gives us enough Madi to supply labour for the Nimule post and still leaves it possible to administer from Kadjo-Kadji.'

On at least one occasion Kelly reveals that his view of the tribes was not always objective but was influenced by sentiments which had nothing to do with administrative convenience:

> It would be a pity for the Sudan not to 'get' the prosperous people of Farajok and Obbo who with their fondness of clothes and such marks of distinction as brass bands would be worth having, but I fail to see at present how we can cut them off from the remaining Acholi.

(27 February)

By far the most serious problem in sorting out the tribes was the deterio- rating relationship between Kelly and Tufnell, largely because of their contrasting attitudes to the tribal question, reflecting perhaps their different personalities and backgrounds. Kelly was an adventurous and athletic man, 6′ 6″ tall and former heavyweight boxing champion of the army. He was immensely popular and appears to have been destined for high rank in the British Army. Commissioned into the Royal Engineers in 1899 he had served for 10 years with the Egyptian Army in Sudan. He had been resident engineer for the construction of Port Sudan, and received Egyptian decorations for this work and as inspector of roads. In his attitude to the local people he appears to have been enlightened and generous. He did all in his power to avoid conflict; he insisted on paying for supplies obtained from villagers. Nevertheless, there is evidence in his diary that he was prepared to be quite forceful in achieving his objec- tives. Tufnell had served in the King's African Rifles for five years before joining the Uganda civil service. Shortly before being appointed to the Boundary Commission he had been promoted to District Commissioner after four years as an Assistant DC. Most recently he had been in charge of Rudolf Province which was a vast and troublesome area in which Tufnell spent much time on operations to prevent tribal raiding and disarming the tribes (Barber 1968: 120–37). His methods had inevi- tably involved the frequent exercise of force and punishment of the recalcitrant. Kelly soon found Tufnell's approach contrasting with his own: on the day he met Tufnell and Lilley he wrote: 'I was not a little surprised to learn the way the Uganda people talked of their dealings with the natives; they appear to think nothing of taking cattle and grain by force wherever they go' (15 January).

As time went on it became increasingly clear that Tufnell did not wish to spend much time working out a proper boundary. He was tired of life in the African bush and was longing for leave. He believed that Kelly should accept his own opinions as to where the tribal territories lay, derived from his experience of northern Uganda, and not become involved in detailed reconnaissance. To make matters worse Kelly was unable to conduct the kind of reconnaissance work he wished because the expedition was short of animal feed largely because the Uganda party had failed to produce the quantities they had promised.

On 5 February Kelly wrote:

> Tufnell was much upset when I said that if we could not come to a definite conclusion based on knowledge and not on supposition regarding the hill Acholi, I must delay ten days or a fortnight in these parts to solve the question. It is unfortunate that he is so anxious to rush the whole affair – as he goes on leave directly it is finished, and I also am beginning to feel that the burden of finding out all the information to settle the boundary is lying with me, in addition to the work of mapping.

Matters did not improve, and ten days later on 16 February, with less than one-third of the boundary fixed by the commission, it was agreed that Tufnell should leave the party:

> He can do no good now in the investigation of the frontier to the east of Madial and being fed up to tears after his 20 months of this country, thinks of nothing but getting on leave, so that he handicaps me very much in any exploration work. We agreed to make out our conclusions as to the boundary after which he will go off to Patelle and on leave. I confess I am glad to get rid of him as he has been little help and has left all the burden of gaining information to me. I am moreover not at all an admirer of the methods of which he is the chief exponent, of dealing with the natives.

In his report to the Governor-General, Kelly simply stated that 'on account of water difficulties the country east of Loruwama had to be visited by the Sudan party alone, which was provided with camel transport' (Sudan Archive 186/1/290).

191

THE RECORD OF THE DIARY: 17 FEBRUARY TO 13 MARCH 1913

This sector of the boundary from Madial to Jebel Mogila, a distance of some 80 miles, was delimited by Kelly. By the time the commission had returned to Madial it had marched 300 miles, an average of 12 miles each day for some 25 days. Before parting company it seems that Tufnell and Kelly had agreed on a boundary 'provisionally' at least as far as Jebel Mogila. It may be significant that Kelly's diary now refers increasingly to prominent physical features which the two men may have picked out as convenient markers. There is less reference to tribes and tribal territories in this part of Kelly's diary, partly no doubt because of the rather sparser population. As it happens, the proposed boundary was a straight line running in a northeasterly direction, which apparently conformed to Kelly's understanding of tribal locations. In much of this sector the line was probably satisfactory as at least one tribal chief conceded:

I sent for the headman from the village which was said to be some distance away and told the other people to bring sheep and a bull. We found they had very exalted ideas as to the value of things and refused to sell a sheep for less than 8 rounds of iron wire; bulls were not forthcoming. The chiefs arrived and I harangued them for some time through Mohamed and Lakuda, probably much of what I said was misinterpreted. I said we had come to see the boundary between tribes so that afterwards there may be peace between them, and that perhaps next year we should send troops to settle here and see that the boundaries were observed. They said they were raided by the Boya, the Karoko (Dodinga) and the Turkana. Their boundary to the west is the Zangaiyetta, while they agreed that a line between Zulia and Mogila would be quite acceptable. They said they had no grain and were hungry. After telling them to bring in grain and sheep, I sent them away.

(7 March)

When Kelly's party reached the vicinity of Jebel Mogila it was decided to turn back. It was clear that the country to the east towards Lake Rudolf could not be crossed easily. They were already suffering badly from thornscrub lacerations. The camels were undernourished and were dying from sleeping sickness. Finding water for men and animals was increasingly difficult. There was also some friction with local tribes who were most reluctant to furnish the commission with suitable guides:

As there was no competition to accompany us as guides, we had to induce two, who appeared to know all about the country, to remain with us as guides, by handcuffing them. This is a regrettable necessity but after being told of guides deserting a few times, are becoming rather hard hearted.

(24 February)

The use of handcuffs proved particularly hard-hearted because the key to the handcuffs was lost, and the unfortunate guides had to accompany the commission all the way back to Madial.

THE BOUNDARY FROM JEBEL MOGILA TO LAKE RUDOLF

The Boundary Commission did not visit any territory east of Jebel Mogila (4° 15′ N, 34° 30′ E). A 'theoretical line' (presumably a straight line) was recommended eastwards to Mount Lubur on Lake Rudolf, 'but if the northern portion of the lake proves navigable, a strip of territory should be reserved to Sudan affording a port on the lake'. Surprisingly, these proposals were contained in a letter to General Wingate dated 17 February 1913 and signed by H.H. Kelly and H.M. Tufnell; in other words the proposals were agreed on the day they parted company, and before Kelly and his party had left Madial to travel eastwards (Sudan Archive 186/1/291–186/1/305). The letter adds that the 'exact limits remain for further consideration when the limits of the Turkana and Dabosa grazing grounds are more accurately known'. When Kelly wrote his final report on 22 April 1913 he had become more confident about the straight-line boundary from Jebel Mogila to Lake Rudolf:

as far as is known the line from north of Mogila to the north of Jebel Lubur and thence to Sanderson Gulf on Lake Rudolf will clear all the grazing grounds formerly occupied by the nomads of this (Turkana) tribe.

(Sudan Archive 186/1/291–186/1/305)

On 21 April 1914 the Sudan–Uganda boundary was formally declared by an Order of the Secretary of State, following a straight line (130 miles long) from Sanderson Gulf to Jebel Mogila, 'or such a line as would leave Uganda the customary grazing grounds of the Turkana tribe' (Brownlie 1979: 1,005). These words have caused considerable legal and diplomatic confusion in subsequent years. In 1926 the Rudolf Province of Uganda was transferred to Kenya so this sector of the

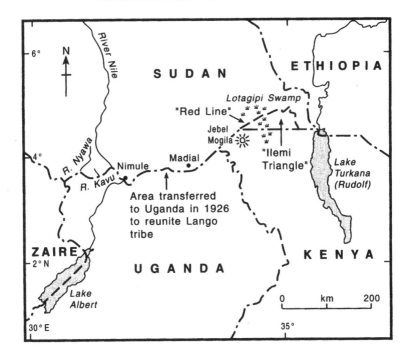

Figure 12.2 The Sudan–Uganda boundary today

former Sudan–Uganda boundary was left to be negotiated with Kenya. In 1931 Sudan and Kenya agreed on the customary grazing lands of the Turkana, delimited by a line known as the Red Line (Figure 12.2). This area is administered by Kenya but there is still some question as to the ultimate sovereignty of the region ('The Ilemi Triangle').

CONCLUSION

To a large extent the present Sudan–Uganda and Sudan–Kenya international boundaries follow the recommendations of the 1913 Boundary Commission whose most influential member was Captain Kelly. The boundary was formally declared by Order of the Secretary of State on 21 April 1914. The boundary was slightly modified by the Secretary of State on 17 September 1926. Part of Jebel Tereitina to the north of Madi Opei was transferred to Uganda to reunite members of the Lango tribe who had been divided by the 1913 boundary (Figure 12.1). Surprisingly, the boundary to the west of the Nile which the commission

194

neither visited nor apparently formally discussed also featured in the final report. It seems that the general alignment had already been agreed in this sector, and all that was required was some suitable physical features to delimit the boundary. The commission does not appear to have been asked to delimit this sector, and the boundary recommended a line which would in any case require 'verification and adjustment on the spot' (Sudan Archive 186/1/291). It is of some interest that part of the boundary was to run 'up the Kayu from its mouth to the Khor Nyaura, thence up the latter to its source' (Sudan Archive 186/1/291). The official description of this sector includes the words 'to the thalweg of the Khor Kayu (Aju) upwards to its intersection with the thalweg of the Khor Nyaura (Kigura)' (Brownlie 1979: 1,005). In reality, the rivers Kayu and Nyaura are not connected. How was such an error introduced into the commission's proposals? A clue lies in Kelly's diary (19 January): 'I had a short talk yesterday ... with an official ... who said ... a suitable boundary would be to follow up the Kayo river and then its tributary the Nyaura to its source ...'

Evaluation of the 1913 boundary and its subsequent political history lies beyond the scope of this chapter. In view of the fact that less than one-third of it was visited and surveyed by the whole commission, it is perhaps surprising how well it has worked. Nevertheless, the work of the commission has been heavily criticised by some authorities, and Kelly's diary helps explain what went wrong better than his official reports to Wingate which speak of the 'great assistance' rendered by Captain Tufnell (McEwen 1971: 257–64; Sudan Archive 186/1/291).

POSTSCRIPT

Following his return to Madial and completion of the work of the commission Kelly embarked on a month of exploration on the Boma plateau near the Ethiopian border. This is also documented in his diary, and includes detail of a tense encounter with an Ethiopian force probing westwards.

NOTE

1 The author acknowledges the kind assistance of Mrs Lesley Forbes, former Keeper of the Sudan Archives at Durham University, who introduced him to the Kelly papers, and Mrs Jane Hogan, currently Keeper, and the help of the British Academy in funding editorial work on Captain Kelly's diary with a view to publication in 1993. Above all, he is greatly indebted to Mrs Elizabeth Pearson of the Department of Geography, Durham University, who

painstakingly deciphered Kelly's appalling handwriting. Without her skill and tenacity over many weeks, the diary would have been of only limited value to the author.

REFERENCES

Barber, J.P. (1968) *Imperial Frontier*, Nairobi: East African Publishing House.

Brownlie, I. (1979) *African Boundaries: A Legal and Diplomatic Encyclopaedia*, London: Hurst/RIIA.

Kelly, Captain H. (1913) 'Diary of the Sudan–Uganda Boundary Commission 1912–1913', Durham: Sudan Archives (unpublished).

McEwen, A.C. (1971) *International Boundaries of East Africa*, Oxford: Oxford University Press.

Nugent, W.V. (1914) 'The geographical results of the Nigeria–Kamerun Boundary Commission of 1912–13', *Geographical Journal*, 43.

Sudan Archive (1913) 'Sudan–Uganda boundary rectification', 17 February 1913, signed by H.H. Kelly and H.M. Tufnell, Sudan Archives, Durham University, 186/1/290.

Sudan Archive (1913) 'Sudan–Uganda boundary rectification: notes regarding proposed boundary line', signed by H.H. Kelly 22 April 1913, Sudan Archives, Durham University, 186/1/291–186/1/305.

INDEX

Abadan 160, 161
Abu Dhabi 18, 175, 180, 181, 182;
   –Dubai war 175
accommodation boundaries 111
Acre 27
acts of war 2
Aden 174, 177
administrative boundaries 73; and borders
   104, 108; see also border
administrative structures 4, 6; civilian 82
Africa 22, 185; the African State chapter 1
agriculture 118, 119, 120, 121, 123,
   173
Aleppo 16, 40
Algeria 7, 9, 22
Algiers Accord (1975) 160, 165, 168
Al-Saud Family 14–16
American University, Beirut 39, 40
Amman 27, 28
Anaza Tribal Confederation 15
Anglo-Belgian boundary 186
Anglo-French partition 129–30
Anglo–Iraqi–Turkish Treaty (1926) 146
Anglo-Ottoman Convention (1913) 146
Anglo-Saudi frontiers 174
Anglo-Turkish Convention (1913) 149,
   151, 153, 155
Aniza 16
annexation 60, 64, 65, 80, 82, 87, 107
Aoun rebellion 51–2, 56
Aqaba 22, 99, 133
Arab: coalition 125; see also Arab League,
   Arab states; independence 131;
   recognition 61; settlements 79–80, 94,
   110, 119; sheikhdoms 163; solidarity
   163 see also Arab nationalism; triangle
   region 94; world 26
Arabia 15, 130, 173; frontiers of 173–4
Arab–Israeli wars xvi, 21–2, 24, 25, 27,
28, 30, 80, 110, 126; peace 68; see also
boundary conflict

Arab–Jewish relations 88, 94; see also
   Arab–Israeli wars, boundary conflict
Arab League 31, 147, 148
Arab nationalism 2, 12, 19, 29
Arab states 13, 16–19, 29, 69
Armenian quarters 43
Armistice lines 76, 102, 103, 107, 108
asabiya 6, 17, 132; see also tribal
Asad regime 23, 24, 26, 28, 29, 30, 31
Asia 22
Assyrian: enclave 43, 47; migration 41
Augsburg compromise 4
authority 12, 14, 18; communal sanction
   of 3; conditional 14; mulk 14;
   paternalistic 17; political 3, 6, 17–18,
   43, 67 see also political legitimacy;
   pre-Islamic 14; religious 16; state 3, 5,
   48, 52; sovereign 8
autonomy 11, 38, 111, 112
Ayatollah Khomeini 11–12
Aziz ʿAbd al-, King 15, 16

Badu tribes see Bedouin
Baghdad 12, 30, 144, 146, 163, 164, 167,
   168, 170; government xviii, 159;
   summit 169; Pact 12
Bahrain 16, 18, 174, 177, 179
Banias River 101, 102, 111
Basra 12, 16, 144, 146, 161, 165, 166
Basta quarter 39
Ba'thist regime 23, 29, 30, 163, 165
Batin wadi 149, 150, 151–2, 156
Bedouin 13, 15, 16, 17, 118, 128, 172,
   176, 177
Beer Sheeba Region 118, 121, 123
Beilin Plan 113, 114; see also Gaza, Gaza
   Strip
Beirut xvi, 22, 23, 25, chapter 3, 121, 125;
   boundaries 35–6; civil war in 35, 36;
   mixed areas of 39–42; religious
   quarters of 37–8, 43–4; urban

development of 39; *see also* East Beirut, West Beirut
Belgium 186; *see also* Anglo-Belgium boundary
Bersan 106
Bethlehem 94
black economy 25; *see also* market economy, public sector economy
blockades 41, 48, 51, 52, 55, 90, 174; *see also* boundary conflict, terrorism
border: administration 182, 186, 190, 191; allocation 132; definition 18, 65; disputes xv, 69, *see also* boundary conflict; settlement 112, 178–9, 180; zone 101, 102; *see also* administrative borders, boundaries, imposed borders, security
boundary: conflict *see* boundary conflict; delimitation of xiii, xiv, xv, xvi, xviii, 10, 50, 65, 99–103, 127, 184–6; inequality 118; marking 184; Muslim xvi; status review 19; urban 37; zone 128–9; *see also* blockades, border, boundaries
boundary agreement (1932) 128, 134–6
Boundary Commission 131, 138, 143, 150, 151, chapter 12; Surveys, problems with 107–8, 109, 133, 134–7, 138, 139, 182, 184–5, 186, 190, 191, 194–5; *see also* British Military Survey
boundary conflict xiii, xvi, xviii, 6, 10, 18, 25, 27, 28–30, 32, 37, 41, 43, 45, 49, 50, 55, 59, 103, 111, chapter 9; resolution of xiii, xiv, 32, 69, 126; *see also* blockades, confrontation lines, civil wars, military intervention, terrorism, violence
boundary sealing *see* boundary closures, sealed boundaries
boundaries: xvi, 47, 71–3, 76–7, 115–21, 130, 138, 172, 194; function of 21, 71–2; permeability of 21, 25, 27, 32, 37, 44, 45, 81 *see also* transboundary flows; realignment of xiv, 128, 139; removal of 77–80; *see also* accommodation boundaries, administrative boundaries, borders, Christian boundaries, confessional boundaries, cultural boundaries, *de facto* boundaries, erased boundaries, ethnic boundaries, felt boundaries, formal boundaries, green line boundaries, imposed boundaries,

international boundaries, Islamic boundaries, municipal boundaries, natural boundaries, new boundaries, open boundaries, perceived boundaries, water boundaries – and those for individual countries
Britain *see* United Kingdom
British colonialism *see* colonialism
British–Egyptian condominium 186
British empire 173; *see also* colonialism
British mandate 174
British imperial policy 13; *see also* colonialism
British Military Survey 129, 130, 132, 134, 135, 136, 151, 152, 153, 155; unsurveyed territory 195; *see also* Boundary Commission
Bubiyan Island 146, 150, 158, 159–63, 164, 165, 166–9, 170; *see also* Warba Island
buffer zones 65, 66
Buraimi Oasis dispute 18, 174, 175, 177, 179, 180, 181
Burrows, Bernard Sir 177, 178, 179

Cairo 31, 114; conference 130
Camp David Agreement 111, 114, 117, 124
capitalism 24
cease-fire line 51, 56, 107, 108
Central Middle East chapter 2
Ceuta 8
check-points 48, 49, 50, 90; *see also* blockades, sealed boundaries
Christian–Muslim boundary 37, 52, 54
Christian quarter 39, 40, 41, 42, 43, 45, 51
citizenship 66, 67, 69
civil wars 22, 23, 24, 25, 41, 42, 45, 46–7, 50; Jordan 27, Lebanon 23, 30
cold war 110, 124
colonialism xv, 4, 8, 9, 10, 11, 14; boundary markers for xviii–xix; British 6; French 6; Israeli 114; *see also* decolonisation
common market 69
communication barrier 73
communism 26
confessional boundaries 36, 38, 40–1, 42–5, 46, 53
confrontation lines 45, 49–50; *see also* boundary conflict, violence
constitutional law 14; Islamic 2, 3, 14; *see also* international law

corruption 25
Costa Rica xiii
cross-boundary interaction xvi, 27, 32, 72; education 79; marriage 79
CSCE 69
CSUCA xiii
cultural boundaries 38
Cyprus 56

Damascus 23, 28, 30, 31, 32, 35, 37, 38, 40
Dbayye 39
Dead Sea 99, 104, 106, 107, 108, 109
decolonisation 7–8, 9; post-colonial period 9
de facto boundaries 53, 55
Deir Yassine Massacre 39; *see also* conflict, terrorism, violence
demarcation lines chapter 3
demilitarised zones 102
democracy 12, 64
depopulation 76; *see also* population
destruction 35; *see also* terrorism
development xvi, 76, 77, 82, 84, 85, 167; capitalist 26; cultural 7; social 7; urban 117, 118; *see also* economic development
Dirah 17, 18
Druze 38, 43, 48; rebellion 140
Dubai 175, 177, 178
Durham University xiii, xiv, xviii, 195
dynastic succession 3, 16; divine right to 14

East Africa 33
East Bank 13; *see also* Israel, West Bank
East Beirut 45, 47, 50, 51, 52, 54
Eastern Arab Front 28, 30
Eastern Bloc 26
East–West divide (Beirut) 35, 38, 42, 45, 52, 53, 56
economic: contract xvii, 71, 73; control 1; complementarity 21; decline 29; dependency 96; equality 62; exploitation 139; liberalisation 26; mismanagement 25, 26; zones 126
economic cooperation 28, 62, 67, 112
economic development 7, 113, 115, 121–3, 125
economic integration 25, 32, 69, 78, 96
education 123
Egypt xvii, 12, 22, 28, 30, 31, 40, 69, 110, 111, 112, 114, 117, 121, 122, 124, 126, 130

Egypt–Israel peace initiative 31, 125; agreement 123, 124; authority 123; negotiations for 113, 123; treaty 118; *see also* peace process
El-Arish 114, 117, 118, 120
elections 63–4, 114; *see also* democracy, self-determination
elective caliphates 3
emigration 78; *see also* exile, migration, refugees
employment 95, 96
empty quarters 174, 180
energy 115; *see also* natural resources
environment 115–17, 119–21
erased boundaries chapter 5; *see also* green line boundaries
Ethiopia 69, 188
ethnic boundaries 5, 8, 43, 50, 90, 92
ethnic relations 73, 94, 95
Euphrates River 16, 29, 31
Europe 1, 4, 10, 21, 22, 41, 43, 71, 77, 96, 110
exile, enforced 63, 64; *see also* migration, refugees
exploitation *see* colonialism, imperialism, tribal

Faisal, King 130, 134
family, state control of 3
fanaticism 35; *see also* religious, terrorism, violence
'felt' boundaries 47
First World War 6, 38, 129, 139, 146, 173
formal boundaries 44, 77, 118
France 7, 23, 39, 99, 129, 130, 174
Franco-British Convention 130–1, 135, 136
frontierisation 76–7, 147, 152, 153, 173; Arabia 173–4; arbitration of 175–7; awards 181–3; western concept of 173
Fujairah 178, 180, 181, 182

Gaza 60, 67, 77, 79, 90, chapter 7; free port 123
Gaza mini-state xvii 116, 121, 122, 123, 124, 125
Gaza Strip xvii, 61, 65, 66, 67, 74, 78, 80, 90, 96, chapter 7; agriculture 119–20; economic prospects 121–3; population 120; solution for 123–6; tourism potential 121–2; Israeli settlements on 119
Gemayyel Bachir 47, 48, 50
Geneva Convention 63, 67

Golan Heights 69, 92, 110, 111, 125
Greek-Catholic quarter 38, 40, 42, 48
Greek-Orthodox territory 37, 38, 42, 44, 48, 55, 56
Green Line boundary xvi, xvii, 35, 46, 47, 48, 51, 53, 56, chapter 5; Beirut 4, 7, 51, 55, *de facto* emergence of 80–92; demarcation of 93–4; development near 84–5; enforcement of 90–2; evolution of 75–93; government policies for 76–7, 92–7; Israel 74, 75, 76, 77, 78; non-existent 87–8; pre-boundary stage 76–7; removal of 77–8, 94; settlement policy for 81–5; violence over 88–90
Gulf *see* Persian Gulf
Gulf of Aqaba (Eilat) 104, 106, 107, 122, 139
Gulf war xv, 26, 31, 59, 65, 90, 110, 169

Hadda Agreement (1925) 133, 134, 135, 136, 137
Haifa 40, 77, 129, 134
Haifa–Baghdad pipeline (oil) 129
Hamasa 175
Hamas movement 124, 125
Hamat Gadar (Alhama) 99, 104, 105, 107
Hamra 40
Hamras-Ras, Beirut sector 41
Hamza line 174
Hashemite monarchy 13
Hebron 77, 87, 94
Hejaz 130, 132
Hermetic boundaries *see* sealed boundaries
Hijaz 16
Hizballah 48, 50
homogeneous areas (Beirut) 36, 41, 44, 55
housing 83–4, 123
Huleh Lake 99, 101, 102
Hussein, King of Jordan 13, 26, 28, 109, 110, 113
Hussein, Saddam 29, 31, 110, 114, 142, 147, 165, 168
hypothetical boundaries *see* erased boundaries

IBRU xiii, xiv, xv, xvii, xviii
identification patrols 49
Ihram 19
Ikhwan movement 15, 132
immigration 28, 108, 110; *see also* emigration, migration, refugees
Imperialism, British 13, 14

Imperial route (east) 128, 130, 133, 138, 139
imposed borders *see* border
imposed boundaries 18, 139, 174
independence 6, 8, 18, 112, 114, 124, 182
independent states 7, 8
India 13, 131; government office of xix, 154, 155, 160, 161, 162
industry 95, 120, 121
instability 10
interaction *see* state; *see also* regional cooperation
internal frontiers 35, 42–5
international boundaries xiii, xiv, 8, 73, 74, 84; *see also* border, boundaries
international law 1, 2, 3, 5, 7, 8, 9, 130, 151
international relations 2, 4
inter-territorial boundaries 54, 56, 72, 74; intra- 55, 73, 94
inter-Trucial state frontiers 175–8; *see also* United Arab Emirates
intifada 68, 78, 80, 85, 87, 88, 90, 110, 112
investment 39–40, 42, 69, 95, 123
invisible boundaries xvi, 43, 53, 73, 87; *see also* permeable boundaries
Iran xviii, 11–12, 13, 29, 31–2, 69, 130, 165, 167, 168; –Iraq relations 158, 163, 169; –Iraq peace agreement 168; –Iraq boundary 170
Iran–Iraq war 22, 26, 29, 32, 147, 166–9
Iraq 2, 12, 13, 14, 22, 26, 41, 59, 69, 110, 111, 114, 124, 130, 132, 134, 137, 139, chapter 10; –Najd boundary 133, 135; sovereignty chapter 9, 160
Iraq–Kuwait conflict xviii, 59, 143, 144–7, 147–9, 156, 162, 163, 170
Iraq–Kuwait land boundary xv, xviii, 132, 135, chapter 9, 162, 164, 166, 167–8, 169, 172
irrigation 108, 119, 120
Islam 19
Islamic: alliance 30; constitutional concepts 10, 17; fundamentalism 28, 125, 126; governments 11–12; movements 10; revolution 11, 165; state 2–3, 7, 13; world 2, 4, 17
Israel xvii, 13, 22, 24, 25, 26, 27, 28, 30, 59, 60, 61, 66, 69, 70, 74, 76, 77, 79, 82, 84, 86, 88, 90, 92, 95, 96, chapter 6, 111, 112, 113, 117, 121, 122, 123, 125, 126; aggression 48, 78; occupation 24, 25, 27, 59, 64, 112;

pre-1967 border 66, 69, 78, 80, 81, 82;
  –Egypt land contribution 124, 125;
  –Sinai border 118; *see also* East Bank,
  Green Line, West Bank
Israeli–Egyptian relations 112, 114
Italy 6, 7, 129

Jabal Shammer 15
Jaffa 76
Jebel Anaza 128, 133, 134, 135, 136, 138,
  139, 140
Jebel Druze 131
Jebel Khurgi 131, 140
Jebel Mogila 192, 193–4
Jebel Tenf 131, 135
Jerusalem 68, 69, 76, 87, 88, 94, 111; East
  60, 65, 80, 94, 125; West 64
Jewish: colonisation 80; settlements 59,
  64, 67–8, 70, 78, 85, 86, 94, 95;
  underground 85
Jordan 13, 14, 22, 41, 66, 69, 70, 74, 77,
  95, 96, 107, 108, 110, 111, 113, 119,
  122, 123, 139; –Iraq boundary 134,
  140; –Iraq–Saudi Arabia boundary
  139; *see also* Transjordan
Judea 111, 112, 114

Kaf 133, 134, 140
Karantina Camp 39
Kelly, Captain chapter 12; diary extracts of
  189–94
Kenya 193, 194
Khaldun Ibn 17
Khors 153–5, 156, 161, 194; Abdullah
  165, 170; Bubiyan 164; Khos 188;
  Zubair 147, 149, 150, 153, 154, 158,
  160, 161, 162, 163, 166, 169; *see also*
  natural boundaries
Kurdish quarter 43, 165
Kurds 39, 41
Kuwait xv, xviii, 2, 13, 15, 18, 22, 114,
  chapter 9, chapter 10, 173; boundary
  demarcation 149–55, 169; conference
  133; resistance 159–63, 164, 165,
  166, 167, 169
Kuwait Bay 16, 29, 59, 159, 160
Kuwait–Iraq boundary 158

labour 69, 78–9; cheap 78, 96; skilled 40
Lado enclave 186
Ladhiqiyah 25
Lake Rudolf 192, 193–4
land 39, 60, 65, 66–7, 76, 79, 109, 114,
  115–21, 122, 169, 173; boundaries

xiv, 159; concessions 161, 175;
  corridors xvii, 119, 122, 123; for peace
  exchange 60, 111, 169; international
  xiii
League of Nations 12, 104, 128, 130, 134,
  138, 146, 160
Lebanese: demarcation 35; forces 47, 48,
  51, 52; republic 56
Lebanon 22, 23, 26, 40, 48, 69, 70, 122,
  125, 130; invasion of 24, 25, 48
legitimisation 3, 4, 10, 14
Levant 6, 23, 24, 37, 40
liberation movements 10
Libya 6, 7, 22
Likud Party 111, 112, 113
London Agreements 113

Madi 191, 192, 193; frontier line 190,
  191; tribe 190
Madrid Conference 126
malik 14
maritime boundaries xiii, 155–6; *see also*
  water boundaries
market economy 53, 77, 78, 95, 96; *see
  also* black economy, public sector
  economy
Maronite Christians 23, 37, 38, 42, 48,
  55; quarter 38, 39, 41, 43, 44; militia
  45
Mecca 13
Mediterranean 22, 29, 30, 31, 75, 76, 78;
  Med–Dead Canal 123
MEEC 69
Melilla 8
mental boundaries 38; *see also* perceived
  boundaries
Middle East xv, 4, 5, 11–19, 124; Central
  chapter 2
migration 39, 41, 42, 79–80; rural–urban
  38, 39, 40, 41, 55; rural 42, 50
military 23, 26, 45, 48, 65, 68, 69, 110,
  112, 113; survey 151, 152, 153, 155
military administration 51, 74, 81, 82, 91
military intervention 18, 27, 29
military occupation 48, 60, 63, 67, 68
militia 12, 41, 46, 113; Christian 45, 47,
  50; local 55, 85; sovereignty of 49–50
mixed communities 8, 9, 38, 40, 41, 42, 71
monarchy 14
Morocco 7, 9, 14, 22
Mosul 12, 144, 146
Mount Hermon 56, 111
Mount Lebanon 38, 39, 40, 41
Muhammarah, Sheikhdom of 11

mulk 14
municipal boundaries 80–2
Muscat 177, 179, 181
Muslim Alliance 30
Muslims 41, 42, 43, 45

Nadji Kingdom 173
Najd 15, 16, 132, 134, 135
Nasser Gamal ʿAbd al- 40, 43
national community 5, 11, 19
nationalism 3–5, 10, 42
National Revolutionary Council 148, 149
nation state, chapter 1, 173; *see also* state
natural boundaries xix, 153, 192; *see also*
  khors, land boundaries, water
  boundaries
natural resources 66–7, 95, 108, 115; gas
  121; water 6, 29, 30–1, 60, 66, 69,
  115, 120; *see also* oil, water
  boundaries
negotiations 110, 112, 124, 125;
  Saudi/British 174; *see also* peace
  process
new boundaries 45–50
Nigerian–Cameroon boundary 184, 186
Nile Delta 114, 120, 121
Nile river 120, 186, 187, 194; –Congo
  watershed 186
North Africa chapter 1

OAU 7, 8
occupied territories 2, 59, 60, 61, 62, 63,
  64, 66, 67, 68, 116; *see also* Israel
oil 4, 6, 13, 17, 18, 29, 30, 31, 40, 42,
  144, 155, 166, 174–5; exports 22, 31
Oman xviii, 14, 16, 17, 173, 174, 175,
  177, 178, 181, 182; frontier chapter 11
open boundaries 71; *see also* boundary
  conflict, cross-boundary flow,
  permeable boundaries, sealed
  boundaries
organisation, political 5; social 5
Ottoman Empire 3, 11, 12, 15, 37, 38,
  118, 128, 129–30, 144, 146, 147, 162,
  173

Pahlavi, Muhammad Reza, Shah 11, 163,
  165
Palestine xvii, 27, 45, 48, chapter 4, 74,
  75, 77, 80–1, 85, 86, 95, 97, 99, 101,
  102, 106, 107, 111, 113, 114, 130;
  problem xv, 61–4, 123
Palestine–Transjordan boundary 104–8
Palestine–Syria boundary 102–3

patron–client lineage patterns 9, 17
peace process xvii, 28, 31, 59, 62, 63–70,
  80–1, 92, 95–6, 111, 113, 114, 123,
  125, 126, 143; *see also* negotiations
pearling 17, 18, 175
perceived boundaries 72, 74, 92
Persian Gulf xviii, 6, 16, 19, 22, 40, 124,
  128, 130, 150, chapter 10; region 11, 13
PLA 112
PLO 59, 63, 64, 68, 110, 112, 124
PNL 47, 63, 64
political authority *see* authority
political boundaries 5, 9, 53, 56, 65, 71,
  72, 75, 78, 84, 85, 92, 93, 94, 95, 96
political conflict 16, chapter 2, 36, 42, 45,
  48, 55, 72, 76, 78; resolution of 95–6,
  118, 119
political equality 62, 65
political identity 9, 53, 65
political legitimacy 4, 5, 94
political structures 3, 7, 11, 17, 43, 51, 52, 67
population xvi, 10, 13, 17, 32, 36, 37, 38,
  39, 40, 43, 45, 62, 66, 76, 85, 86, 90,
  124; Arab 111, 113, 115–16, 124;
  displacement 59, 64; exchange xvii;
  growth 37, 115–17; movement xvii, 9,
  21, 23–4, 31, 35, 37, 38, 70, 78, 114,
  115; resettlement 126; rural 9,
  suburban 55; urban 9; *see also*
  migration, refugees
post-colonial: period 9; state 4; *see also*
  colonialism, independence
poverty 41, 42, 43, 48, 124, 175
public sector economy 26; *see also* black
  economy, market economy

qabadays 49; *see also* militia
Qadhafi, Muammar al- 22
Qatar 16, 18, 174
Qatif 15, 119
Qawasim 16

Rabat Arab League Conference 63
Rafia 117, 118, 120
Raggad Gorge 111
rapprochement 22, 28, 29, 30, 31; *see also*
  peace process
Ras-al-Khaimah 176, 177, 178, 179, 181,
  182
Ras Beirut 43, 45
reciprocity 62, 65
refugees 13, 24, 27, 39, 40, 41, 59, 60, 63,
  65, 114, 115, 122, 123; right of return
  66

regional cooperation xvi, 17, 21, 32, 33, 62, 69, 139
regional isolation 30
religious: affiliations 41, 53; fundamentalism 124; legitimation 18; loyalty 40–1; purification 14; sects 42; spatial differences 39; status 15; territories 43, 45; variations 53, 54, 55, 56; unity 3
repression 6, 12
re-unification (Beirut) 50–3, 118
rights 4, 14, 18, 63, 66, 69, 102, 119, 174, 178; inalienable 62–3, 119; of Usufruct 102, 103, 108; see also sovereignty
River Jordan 75, 76, 78, 94, 99, 101, 102, 104, 106, 109
Riyadh 15; line 174
Rue de Damas 35, 39, 40, 41, 45
rural coalition 17

Sadat, Anwar 31
Safwan 149, 150, 152–5, 156
San Remo Conference 130
Samaria 111, 112, 114
Sa'ud Ibn 14–15, 128, 130, 131, 132, 133, 134, 135, 138, 139, 140, 173; see also Wahhabi movement
Saudi Arabia 11, 13–16, 18, 28, 31, 40, 128, 137, 139, 174, 177
sealed boundaries xvi, xvii, 22, 38, 43, 55, 71, 72, 76, 77, 90, 92, 95, 96, 112; Syria 31; Syria–Iraq 29; Syria–Jordan 27; see also border conflict
Sea of Galilee 99, 101, 102, 103, 104
Second World War 21, 153, 163, 174
security 60, 65–6, 68, 113, 143, 175
self determination 7, 8, 9, 60, 62–3, 113, 114, 125
shar'ia 2–3, 4, 11
Sharjah 178, 179, 182
Shatt-al-Arab xviii, 159, 160, 165, 168, 169, 170; boundary 162; crisis 163–6; dispute 158, 169; see also Baghdad Summit
sheikhdoms xviii
Shiite 32, 39, 41, 43, 44, 55; influx 48; militia 48; Amal movement 48, 50
Sidon 37
Sinai 114, 117, 118, 120, 124; Canal 120
slave trade 175
souks 37, 39
sovereignty xviii, 1, 3, 8, 10, 12, 13, 14, 16, 18, 68, 76, 113, 126, chapter 9;

communal 5; disputes over xv; Islamic 2, 17; state 1–2, 11–19, 60; religious 16; rights 4; territorial 5, 10, 11, 12–13, 14, 18, 19, 80, 173; see also authority, rights
Soviet Jews 59, 110
Soviet Union (former) 73
Spain 7, 8
state: chapter 1, 63, 70, 96; boundaries 64–5, 71, 72; centres 78; development 10–11; see also development; interaction xvii 21, 23–6, 75, 78–9 see also economic cooperation; land-locked 96; non-European 5–6; socialist Zionist 111; sovereignty 11–19 see also sovereignty; structures 12; two-state option xvi–xvii, 60–1
succession, principles of 3
Sudan xviii, 186, 187, 193; archive 191, 193, 195; –Kenya boundary 194
Sudan–Uganda Boundary Commission (1913) chapter 12
Suez Canal 22, 114, 118, 119
sulta 14
Sultan of Muscat 174, 177, 178, 179, 180
Sultinate of Muscat xviii
Sunni quarters 37, 38, 39, 41, 43, 44, 45, 48, 55
Sykes–Picot Agreement 130
Syria xv, 1, 12, 16, chapter 2, 35, 40, 41, 42, 52, 69, 70, 101, 110, 111, 130, 131, 132, 135, 140; interaction with Iran 31–2; interaction with Iraq 29–31; interaction with Jordan 26–9; interaction with Lebanon 23–6, 60
Syria–Transjordan Boundary Commission 131, 138, 140

Tartus 25
Teheran 31; –Baghdad discord 165; Treaty (1937), 160, 163
Tel Aviv 95, 122
territorial adjustments 50, 65, 75, 93, 111, 161, 166, 181, 186
territorial control 10, 17, 48, 50, 79, 132
territorial definition xvii, 18, 65, 70, 178–9
territorial disputes xviii, 7, 15, 19, 42, 62, 73, 92, 96, 128; see also blockades, boundary conflict, terrorism, violence
territorial exchange 93–5, 139
territorial limits xv, xvi, xvii, 2, 11, 53; see also Green Line boundary
territorial sovereignty 1, 8–9, 14, 17,

106–7; *see also* sovereignty
territory, chapter 1, chapter 10, 178–9,
    187, 195; colonial 9; delimitation xiv,
    8, 47; mandated 130; merged 67;
    microscopic 50; military 48; national
    7; neutral 39
terrorism 6, 28, 31, 35, 36, 112, 124
thalweg (as boundaries) 151, 152, 155,
    156, 160, 163, 165, 168, 169, 195; *see
    also* natural boundaries, water
    boundaries
tourism 32, 121, 122, 125
trade xvi, 13, 17, 21, 22, 26, 27, 28, 29,
    31, 32, 33, 69, 75, 121, 132, 174–5;
    barriers 73; routes 22, 130; systems 17
transboundary: cooperation xiii, xvi,
    chapter 2; flows 21, 22
Transjordan 75, 104, 106, 107, 130, 131,
    132, 133, 134, 137
Transjordan–Iraq boundary xvii, 135,
    chapter 8; after 1932 136–8
transport 75, 77, 82
travel 22, 24, 27, 30, 87
Treaty of al-Muhammerah 132
tribal: boundaries 186, 188, 189; conflict
    xviii, 188, 190; exploitation 17;
    solidarity 132; structures 11, 17, 18;
    territories 192
Tripoli 25, 37
Trucial Sheikhdoms *see* United Arab
    Emirates
Tufnell, Captain 186, 188, 189, 190, 193,
    195
Tunisia 7
Turkey 22, 31, 41, 69, 144

Uganda 186, 187
umma 2, 6, 14
Umm al-Qaiwain 176, 177, 178, 182
Umm Qasr 147, 149, 153, 154, 155, 158,
    161, 163, 164, 165, 166
unification 31
UNIKBDC 143, 153, 155, 156, 169
United Arab Emirates (UAE) xviii, 16, 18,
    174, 175, 177, 178, 179–81, chapter 11
United Kingdom xvii, xviii, 12, 17, 18, 23,
    59, 99, 107, 128, 129, 130, 132, 144,
    145, 146, 150, 151, 186
United Nations xv, xviii, 1, 6, 66, 111,
    114, 144, 145, 149, 171; Charter 2, 7,
    9, 62, 144; observers 56;
    peace-keeping force 65; Revolutionary
    Council 150; Security Council 2, 144
UN Security Council Resolutions 30,

64–5, 66, 143, 144, 149, 150, 151,
    152, 167, 169
UNWRA 63, 115, 120
USA 21, 59, 66, 70, 110, 113
Uqair Protocols 132, 133, 134, 135, 136,
    174
urban–non urban boundary 37;
    deconstruction of 51, 55
uti possidetis juris 8–9, 13
Uymaq 11

velayat-e-faqih 11, 12
Vienna Convention, Law of Treatise 1969
    (Article 46) 148–9, 152
violence 4, 6, 35, 36, 39, 45, 51–2, 85; *see
    also* boundary conflict, terrorism

Wadi Abou-Jamil 41
Wadi al-Taym 37
Wadi Arabah 104, 106
Wadi el Azariq 117
Wadi Hadf 181
Wadi Hanifa 15
Wadi Hatta 181
Wadi Madha 179, 181
Wadi Qor 182
Wadi Sirhan 133
Wahhab, Muhammad Ibn ʿabd al- Sheikh
    15; *see also* Wahhabi Movement
Wahhabi Movement 14, 15, 132, 139
war, internal 45–50
Warba Island 145, 146, 148, 150, 158,
    164, 165, 166–9, 170; Iraqi demands
    for 159–63; leasing of 168, 169;
    cessation of 169; *see also* Bubiyan
    Island
water *see* natural resources
water boundaries xvii, chapter 6, 162; lake
    chapter 6; river chapter 6, 151;
    problem with 104–9; sea xiv, 126,
    150, 175
West, the xv, 17, 26, 173
West Bank xvii, 13, 27, 60, 61, 65, 66, 67,
    74, 76, 77, 78, 79, 80, 81, 82, 84, 85,
    87, 90, 95, 96, 99, 107, 111, 113, 114,
    125, 126; Arabs 111; citizenship 119;
    settlements 84, 94
West Beirut 42, 45, 47, 48, 50, 52, 53, 54

Yarmuq river 99, 104, 107, 108, 109, 111
Yemen 16, 173, 174

Zaʾim system 43, 49
zakat 18, 176, 178

Printed and bound by CPI Group (UK) Ltd, Croydon, CR0 4YY

01/11/2024

01782616-0006